Benjamin Knox Rachford

**Some Physiological Factors of the Neuroses of Childhood**

Benjamin Knox Rachford

**Some Physiological Factors of the Neuroses of Childhood**

ISBN/EAN: 9783337372866

Printed in Europe, USA, Canada, Australia, Japan

Cover: Foto ©berggeist007 / pixelio.de

More available books at **www.hansebooks.com**

# SOME

# PHYSIOLOGICAL FACTORS

OF THE

# NEUROSES OF CHILDHOOD.

BY

## B. K. RACHFORD, M.D.

Professor of Physiology and Clinician to Children's Clinic, Medical College of Ohio; Member of Association of American Physicians and American Pediatric Society, etc.

CINCINNATI:

THE ROBERT CLARKE COMPANY.

1895.

# PREFACE.

This little book is for the most part a republication of a series of papers first published in the *Archives of Pediatrics.*

In preparing these papers for republication I have thoroughly revised them and made many additions. The chapter on auto-intoxication has been entirely rewritten, so that from an unpretentious paragraph it has become the most important chapter in the book.

B. K. Rachford, M.D.

Cincinnati, *Sept.* 1, 1895.

# CONTENTS.

(v)

# CHAPTER III.

# CHAPTER IV.

# CHAPTER V.

## CHAPTER VI.

### VENOUS CONDITION OF THE BLOOD.

## CHAPTER VII.

### AN IMPOVERISHED CONDITION OF THE BLOOD.

## CHAPTER VIII.

### REFLEX IRRITATION.

## CHAPTER IX.

### EXCESSIVE NERVE ACTIVITY.

# SOME PHYSIOLOGICAL FACTORS

# NEUROSES OF CHILDHOOD.

## I.

### NORMAL FUNCTIONS OF NERVE CELLS.

The term " Neuroses of Childhood " is here used
to cover all local and general nervous disorders
which do not depend on known local pathological
lesions of the nervous system. This definition of
the term neuroses does not imply that these diseases
have an entirely unknown pathology, but only that
they can not be morphologically classified. In
these diseases we know more of the symptoms than
we do of the lesions, more of the effect than we do
of the cause, more about the disordered functions
of nerve cells than we do of the widely varying pathological conditions which produce these
disordered functions, and this is the reason why
these diseases are incorrectly called functional
nervous diseases.

(1)

The first requisite to the study of the abnormal functions of nerve cells should be a knowledge of the normal functions of nerve cells. For this reason the following preliminary physiological outline is introduced.

Nerve cells have three important functions, viz.: to generate, to discharge and to inhibit energy.

**The highest function of the nerve cell is to generate energy.** By this is meant that the cell transforms and appropriates existing energy. The amount of existing energy is constant, the cell does not and can not originate energy, but in the chemical metabolism necessary to the life of the cell force is developed which is transformed into that form of nerve energy which is the special function of the individual cell (Prof. J. Gad—personal communication), and this nerve energy is stored up to be discharged in the exercise of the cell's peculiar function. From this it would follow that the generation of nerve force would be directly dependent on the healthful chemical metabolism of the nerve cell, but it does not follow that the *amount* of energy thus developed would always be commensurable with the physical waste or the chemical metabolism going on in the cell. This disproportion between cell activity and the amount of force developed is especially noticeable in *the immature nerve cells of the child.* A most marked example of the slight amount of energy developed by the cell activity of immature cells may be noted in cortical cells of the brain of the infant, and the brain of the unintelligent adult. In such brains the cortical cells concerned in the development of mental energy have going on within them an active chemical metabolism with the development of very little mental energy, and this failure of chemical metabolism to develop commensurate mental energy is due to the incomplete functional development of these cells. Of all the cells in the body the cell that develops mental energy is the slowest in reaching the degree of functional perfection for which it is destined, and it only does so after a judicious training in the exercise of its peculiar function throughout a long period of about twenty-three years.

The functional development of the motor cell is much more rapid, and the disproportion between the amount of cell activity and force produced is not so great as in the mental cell, but nevertheless it may be stated as a fact true for all nerve cells that the amount of energy which a cell is capable of generating will depend on the degree of functional development which the cell has attained.

But these facts concerning the difference in the amount of cell energy developed by different cells under the same conditions do not in any way modify the force of the statement made above that nerve energy is directly dependent on the chemical metabolism of the nerve cell.   It will therefore be permissible for us to say that other conditions being the same, *the amount of energy developed by a nerve cell will directly depend on the amount of healthful chemical metabolism going on within it.*  This point in the physiology of the development of cell energy is very important, since upon it rests the conclusion that insufficient nourishment will diminish the capacity of the nerve cell for the generation of energy.   The maximum amount of energy will therefore be found stored in the well nourished cell, and the minimum amount of energy in the starved cell.   We shall see later that this statement which has important clinical bearing can be strongly supported by experimental evidence.

**Discharge of nerve energy** is a function of the nerve cell only second in importance to the generation of energy.   The more or less constant discharge of force is an automatic function of the nerve cell, and this unconscious discharge of nerve impulse is the regulating function that controls the whole body mechanism. As an example of this automatic discharge of nerve force one may cite the influence of the central nervous system over involuntary muscular tissue, whereby the "muscular tone" of involuntary muscles is maintained, the vaso-motor center in the medulla oblongata has such an influence on the muscular coats of blood-vessels as to keep them in a state of normal contraction best adapted for the purposes which they serve, this vascular tone remains much the same at all times except when the functions of the center are perverted by some change in the metabolism of the cells or by influences acting on the center either directly or in a reflex manner. But possibly of even greater importance to us in this study is the tonic influence of the spinal motor cells on the sphincter muscles of the stomach, the anus and the bladder, all of which are dependent on the spinal cord for their normal muscular tone. The "muscular tone" of these sphincter muscles is easily disturbed by reflex stimulation, producing on the one hand spasmodic stricture and on the other incontinence. The muscular tone of the skeletal muscles is likewise said to be maintained by an automatic discharge of nerve force, and a perversion of this function may in the same man-

ner produce complete relaxation or irregular spasmodic contraction of these muscles. These examples on the part of the muscles are sufficient to
illustrate how nerve cells, by the automatic discharge of nerve force regulate the whole body
mechanism. It would be of no value for us to dis-
cuss whether this more or less constant discharge
of nerve force is purely an automatic function
of the cell or whether it is due to unconscious
afferent impulses producing a reflex discharge of
force. It is sufficient for us to know that these
phenomena exist and it is a matter of words
whether we speak of them as automatic or as
reflex.

Nerve force may also be discharged *voluntarily.*
This power of willing the discharge of nerve impulses resides in the cortical cells of the cerebrum.
The influence of the will over the discharge of
force, by the spinal motor cells, is a physiological
fact of great clinical importance in the study of the
neuroses of childhood.

Thirdly and lastly and most important of all, so
far as our present study is concerned, *nerve force
may be discharged reflexly;* this reflex discharge of
force occurs when nerve cells are acted on by outside stimuli. If the stimulus be mild the reflex

discharge of energy from the normal motor cells
of the cord occurs only through the paths of least
resistance, viz.: the afferent nerves in the same
plane and on the same side as the nerve fiber that
carried the afferent stimulus, but if the stimulus be
more severe the reflex discharge of force will also
occur in the same plane but on both sides of the
cord. We shall see later how these simple laws of
reflex action have little control over the reflex discharge of nerve force under certain pathological
conditions.

**Inhibition of nerve energy** is the third important
function of the nerve cell. Certain cells throughout the central nervous system have the power of
inhibiting energy discharged by other cells and it
is also possible that some cells of high functional
development may have the power of inhibiting
their own energy. But however this may be it is a
well established fact that inhibition does exist and
that this power of inhibiting nerve energy may be
either voluntary or involuntary. Voluntary inhibition of mental and motor force is a function peculiar to the cells of the cerebral cortex, but involuntary inhibition of nerve force is a function of
cells every-where distributed throughout the central nervous system, but the higher centers are

always the predominating centers when the nervous system is intact. The spinal cord contains cells or collections of cells (centers), which are capable of being excited reflexly so as to give motor expression to sensory stimulation, and inhibition can best be understood by studying the inhibitory influence of the higher centers on spinal reflex acts. The spinal reflex centers can act quite independently of higher centers. Gad demonstrated that after section of the spinal cord at any point, the centers below the section are still active and capable of translating sensory impressions into motor acts. But this absolute autonomy of the spinal reflex centers does not exist when the spinal cord is in normal communication with the brain, then the reflex centers in the cord are more or less under control of other centers higher up in the cord, the medulla oblongata and the brain. These centers may influence the lower spinal centers not only in causing them to discharge force as we have above noted but also in inhibiting their reflex acts which are discharged from any cause whatsoever. Some of the inhibitory influences coming from the brain are voluntary and probably originate in the cells of the cerebral cortex, for example we can by voluntary inhibition control the urinary bladder reflexes and prevent urination even when the micturition center in the lumbar cord is strongly stimulated, and again there are spinal reflexes over which voluntary inhibition has no control, as for example erection, ejaculation, and movements of the iris.

Of even more importance to us in the study of the neuroses of childhood are the involuntary inhibitory centers which are distributed throughout the central nervous system. They are found in the brain, the medulla oblongata, and the spinal cord; and without voluntary effort or apparent reflex stimulation these centers seem to exert a constant inhibitory influence on the lower spinal centers. Setchenow's inhibitory center in the medulla oblongata is an example of similar centers which we have reason to believe exist in the large ganglia at the base of the brain. The inhibitory influence of this center on spinal reflex acts has been quite satisfactorily demonstrated. It is also easy to demonstrate in a brainless frog that stimulation of the sciatic nerve will inhibit spinal reflex acts. It is clear therefore that spinal inhibition may be brought about by other impulses than those that come from predominating centers in the brain and medulla oblongata, that is to say by impulses which are not in themselves of a specifically inhibitory nature, but originate in the peripheral stimulation of sensory nerves. But it is not necessary for me to narrate experiments bearing on the subject of inhibition of nerve force, for such experiments are so satisfactorily detailed in the physiologies that I need here only say that experimental physiology teaches us to believe that there are cells every-where distributed throughout the central nervous system, which have the power of inhibiting nervous energy. It matters little to us in the present study, whether this inhibition is always the special function of certain cells or whether it may also be the function of the nucleus of the cell that discharges the energy; but it is important for us to know that inhibition exists both for mental and motor acts, and it will appear later why a clear understanding of the influences that control and disturb inhibition is of the utmost importance to us in the study of the neuroses of childhood. If kept in mind, the above outline of the normal functions of the mature nerve cell will materially assist in the study of the functional peculiarities which are manifested by the immature cells of the rapidly developing nervous system of the child.

## II.

PHYSIOLOGICAL PECULIARITIES OF THE NERVOUS SYSTEM
OF INFANCY AND CHILDHOOD.

If we turn to our text-books on physiology or diseases of children to inquire into the peculiarities of the nervous system of infancy and childhood, we shall close them with the feeling that very little is known of this important subject. While unfortunately this conclusion is for the most part true, yet we are not so wholly devoid of knowledge on this subject as our text-books might lead us to infer. We have at least some knowledge of a few of the physiological peculiarities of the immature nervous system of the child that have a most important etiological import in the study of the neuroses of childhood, and it is my purpose here to outline such of these peculiarities as I believe to have a bearing on neurotic disease. At birth the brain is morphologically and functionally the most immature of all the great organs of the body. From birth up to seven years of age it develops enormously in weight, in structure, and in func-

tion.   At this time the brain has attained ninety
per cent of its maximum weight (Boyd), and after
this slowly increases in weight up to the age of
eighteen, but increase of function does not keep
pace with increase of weight, the brain of a child
of eight is almost as large as the brain of an adult,
but as Clouston aptly says, " the difference between
what the brain of a child of eight and the brain of
a man of twenty-five can do and can resist is quite
indescribable.   The organ at these two periods
might belong to two different species of animals so
far as its essential qualities go."   While the rapid
increase in weight of the brain does not continue
after the seventh year, the rapid increase in the
brain's functional development goes on and still
continues long after the brain at eighteen has
reached its maximum weight.   Clouston says " the
unique fact about the nerve cell is the extreme
slowness with which it develops function after its
full bulk has been attained.   *   *   In this it differs
from any and every other tissue.   *   *   We may say
that after most of the nerve cells of the brain have
attained their proper shape and full size, it takes
them the enormous time of eighteen or nineteen
years to attain such functional perfection as they
are to arrive at."   It is an important fact that

should always be kept in mind that the entire nervous system of the normal infant and child is constantly undergoing structural and functional development, and that the structural development, so
far as we are now able to judge by our instruments
of precision, is much more rapid than is the development of function. It is also a fact that even
with normal children, this development of structure
and function does not always go on with the same
rapidity, nor does it always follow a regular order
in its development. It is quite within the limits
of health that certain functions may be rapidly developed and that other functions may be unusually
retarded in development. The innumerable conditions of heredity and environment have their influence on the nervous system of the child in developing and retarding both structure and function,
and this interference with the order of development
is not an abnormal condition if within a reasonable
time the delayed functions reach a normal state of
development. But it is not my purpose to enter
deeply into this phase of my subject. I only wish
here to call attention to the following important
facts :

1st. At birth the nervous system is structurally
but more especially functionally immature.

2d. Throughout infancy and the earlier years of childhood the brain normally undergoes rapid structural development.

3d. Throughout the entire period of infancy and childhood the brain normally undergoes rapid functional development.

4th. Innumerable conditions of heredity and environment have much to do with the rapidity and the order of development of the functions of the nervous system of the normal child, as well as with the failure and retardation of their development in the abnormal child.

5th. The metabolism in the normal immature nerve cell of the child must be rapid enough not only to supply the functional waste, but also to supply the material for the growth and development of cells.

6th. The structural instability of the functionally weak and immature nerve cell of the child makes it much more irritable and excitable than the stable mature nerve cell of the adult.

With these general considerations of some of the functional peculiarities of the nervous system of childhood, let us pass to the consideration of certain special functions of the nervous system, which are not the same in childhood as in adult life.

The feeble inhibition of nerve energy is from a clinical stand-point the most important physiological peculiarity of the nervous system of infancy and childhood. The inhibitory function of the nerve cell is the last to be developed; the cell first acquires the function of generating energy, then the function of discharging energy, and lastly the function of inhibiting or co-ordinating energy. These functions of the cell are developed in the order in which they are needed. Until energy is generated there can be no occasion for a discharging function, and until energy is discharged there can be no occasion for an inhibiting function. *Feeble inhibition* is therefore one of the physiological characteristics of the immature nervous system of infancy and childhood, and it plays a most important rôle as a predisposing factor of the neuroses of childhood. Otto Soltsman noted that inhibition was very feeble in young animals, and that it became stronger as the animal got older. The inhibitory function of cells is therefore in this regard like the generating function, it gradually becomes stronger as the cells get older up to the time when they reach their complete functional development. But it must be remembered that the inhibitory function of a cell is always developed later than

that function of the cell which generates the force
which is to be inhibited. In the normal order of
things the function of inhibition should closely follow the development of the function which generates the force to be inhibited.

The inhibitory mechanisms which control the
discharge of nerve force that regulates such vital
processes as the action of the heart and the lungs
are fairly well developed at birth, while those that
regulate reflex phenomena are slowly developed
during infancy and early childhood, and voluntary
inhibition of motor and *mental* force does not find
its complete development till long after childhood
has passed. The late development of the function
of inhibition is a fact of prime importance from a
clinical stand-point, because this is the last function
of the cell to develop and is the one that is most
likely to be still further retarded in development
by unfavorable conditions of heredity and environment. *It is therefore the abnormally feeble inhibition
which occurs in the abnormal child that is such a potent
factor in the production of neurotic disease in infancy
and childhood.* It is my belief that this functional
immaturity of the inhibitory centers is most important in explaining the manner in which childhood acts as a predisposing cause of such reflex

neuroses as convulsions and incontinence of urine. It is evident that this cause, most potent at birth, gradually grows less as the child grows older. This is especially true of voluntary inhibition. At birth voluntary inhibition, if it exists at all, must be very feeble, and it gradually grows stronger as the higher functions of the brain are more and more developed. We have a good example of voluntary inhibition in the influence of the will over urination. One wills to urinate and the impulse passes down the cord to the lumbar center that presides over urination, and it is there translated into the reflex act of micturition, or on the other hand one wills not to urinate and the impulse travels down the cord to the lumbar center, and the act of urination is inhibited.

But the functional immaturity of the *involuntary* centers is of even more importance to us as clinicians than the voluntary, for these centers have most to do with co-ordinating and regulating spinal reflex movements, the lack of inhibition on the part of these centers would make it possible for an overflow of spinal reflex movements to occur and in this way predispose to such convulsive disorders as eclampsia, chorea and epilepsy. As previously noted the reflex discharge of energy from the

spinal motor cells occurs normally through the
paths of least resistance, that is in the same plane
and on the same side, or in the same plane and on
the opposite side of the cord from where the nerve
fiber entered that carried the afferent stimulus.
But if the resistance to the spreading of the reflexes
up and down the cord be reduced, or if the exciting stimulus be increased, then we may have an
overflow of energy up and down the cord exciting
general spinal reflex movements.  As above stated,
these spinal reflex movements are inhibited and an
"overflow" of energy prevented by the action of
involuntary inhibitory centers higher up in the
cord, the medulla oblongata and the brain.  The
normally feeble inhibition of infancy predisposes to
such an "overflow" of spinal reflexes, or, in other
words, to convulsive disorders of all muscles ope-
rated through spinal motor nerves.  It is also easy
to understand how unfavorable circumstances of
environment and heredity, having their greatest retarding influence on the development of the inhibitory function of the immature nerve cells of the
infant and child, would still further predispose to
overflow of spinal reflexes and in this way to convulsive disorders.  By this *overflow* of nerve energy
we may have a large number of spinal reflex move-

ments as the result of a single exciting stimulus. Extensive convulsive movements of almost the entire body may in this way be caused by some simple discharging stimulus. It is one of the functions of the reflex inhibiting mechanisms to prevent this overflow, so that an impulse sent to one portion of the cord may not overflow and spread to other portions of the cord, but the mechanism being inefficient the inco-ordinated and spasmodic muscular movements occur. This overflow of nerve force is not peculiar to spinal cells exhibiting motor energy, but it also occurs in the cortical cells exhibiting mental energy (insanity). Inhibition against this overflow is quite as important in the brain cortex as in the spinal cord.

It is of clinical importance that we should here note that, not only are the reflex centers in the gray or sensory portion of the cord, but the conducting fibers by which reflex movements overflow and spread up or down the cord are also in the sensory tract of the cord, for this gives us a partial explanation of how certain drugs such as cimicifuga, the bromides, antipyrin, and gelsemium, by depressing the sensory tracts of the cord can control reflex spinal movements. But it must be remembered that these drugs given in this way are given

to relieve symptoms and do not have a curative influence by removing the cause of the disease.   In
this connection I may quote Lauder Brunton, who
says that "spasm is as a rule due to diminished
action of the co-ordinating or inhibitory centers,
rather than to excess of action in the motor centers," and "those drugs which *stimulate* the circulation and increase the nutrition of the higher nerve
centers and the co-ordinating power tend to prevent spasm."   In this we have an explanation of
the benefit derived from nitro-glycerine in certain
nervous conditions where the circulation is feeble
and the malnutrition great.

   In the light of the influence of feeble inhibition
as a factor in the production of spasm and other
neuroses, one notorious fact demands explanation,
and that is that the first half year of life, when inhibition is most feeble, convulsive disorders are
least frequent.   There are a number of reasons why
this is so.   The most important is that the motor
areas of newly born animals are not so sensitive
and do not so readily respond to reflex or direct
irritation as in older animals.   Another reason is
that the nervous system of the nursing child is not
so frequently excited by reflex or direct irritants as
the child that is fed on a mixed diet. .

Lack of sensitiveness of the motor areas in infants has an important bearing on the study of reflex neuroses. The motor areas of the nervous system of the newly born do not respond to electrical or other stimuli as readily as in older animals. Purely reflex neuroses are therefore very uncommon in the very young infant. There is at this time in life very little need for the inhibitory control of spinal reflex acts by higher nerve centers, because the spinal reflex centers are so functionally immature that we get but a minimum reflex from a maximum excitation. The feeble inhibition of early infancy is for this reason not so potent a factor in producing disease as it is a little later on, when inhibition is found not to have kept pace with the development of cell excitability. The reflex centers very early in life take on their normal irritability while the inhibitory function is very slow in reaching full development. In this way feeble inhibition comes to play an important rôle in the production of the neuroses of childhood. This lack of sensitiveness of motor centers in the young infant has yet another important bearing, since it is in great part responsible for the *lack of tone* of the sphincter muscles of the infant.

I have previously noted that the muscular tone

of the sphincters was maintained by an automatic
function of the central nervous system, so closely
analogous to reflex action that it seems a difference
of name rather than of function.  Now these reflex
or automatic functions of the cord are so immature
in the newly born that there is a lack of tone of all
sphincter muscles, that is to say an absence of the
normal amount of contraction which afterward fits
them for the purposes they are to serve, and which
depends in great part upon the action of normal
reflex centers in the cord.  This lack of sensitiveness of the reflex centers of the cord in the infant
is in my opinion, a most important factor in producing the incontinence which is characteristic of infantile sphincters.  The incontinence of infantile
sphincters passes away with the functional development of the centers whose function it is to maintain in them the normal amount of muscular tone
that fits them for the purposes they are to serve.
Abnormal conditions of heredity and environment
may much delay the functional development of
these centers and for this reason a complete or partial incontinence may continue long after the period when it should normally disappear.  During
this period, when involuntary inhibition is so
feeble, voluntary inhibition is of great service in

preventing, as it usually does, the diurnal incontinence. But at night when the will is asleep a
minimum reflex will overcome the feeble involuntary inhibition and cause a relaxation of the
sphincters. Besides this any abnormal conditions
of heredity or environment which increase the irritability of these reflex centers will also make it
possible for slight reflex causes to disturb the
" muscular tone " of sphincters and cause either
spasmodic stricture or incontinence. The pathological conditions therefore which produce feeble
inhibition and excitable nerve centers are sufficient
explanation, for the not infrequent condition of incontinence of sphincters during childhood, and it
is not necessary to invoke a cause which does not
as a rule exist, viz., insufficient muscular development.

# III.

It is a well-known fact that children are more prone to fever than adults, and it is also well known that the temperature is more variable in the fevers of infancy and childhood than it is in the fevers of adults. Why this is so, is a question which we now wish to study from a physiologic stand-point. But first let us clearly understand what we mean by the terms high temperature and fever.

By high temperature is meant an increase of the body heat, whether it be due to increased heat production or diminished heat dissipation. When high temperature is due to increased heat production it is a symptom of fever, but when it is due to diminished heat dissipation it is not a symptom of fever.

By fever is meant an abnormal increase of those tissue changes by which the normal heat of the body is produced, that is to say an abnormal increase of the chemic changes which result in disorganizing tissues and breaking them up into car-

bonic acid, water, urea and other products of retrograde metamorphosis.

The fever process is characterized by a chain of
symptoms with which every clinician is familiar,
the most characteristic of these symptoms is increase of body temperature. But it must be remembered that the height of the body temperature
does not always mark the severity of the fever process, and that even a severe and wasting fever may
exist with a subnormal temperature. One may
note at least two reasons why the temperature is
not an index of the severity of the fever process.
First: increased heat production is but one of the
symptoms of fever, which is ordinarily but not necessarily produced by the same causes that produce
fever. Second : even should heat production keep
pace with the severity of the fever process, heat
dissipation may be so rapid or so variable that the
body heat at any given time would not be an index
of the fever process. With this understanding, the
terms fever and temperature will be used as above
defined, and we can proceed to study the influence
of the nervous system on these processes.

Increased tissue metabolism, which is the one
great cause of increased heat production (fever), is
under the direct control of the nervous system, and

the centers which control this metabolism and indirectly the production of body heat are called heat centers.* Certain of these heat centers have the function of discharging force which will increase tissue metabolism and thereby increase the body heat, they are for this reason called *thermogenic* centers.

Other so-called heat centers have the power of inhibiting or controlling the discharge of force from the thermogenic centers, and they are for this reason called *thermo-inhibitory* centers. These thermo-inhibitory centers have no direct influence over the processes whereby the body heat is produced; yet they are of the greatest clinical importance because of their control over the thermogenic centers.

The thermogenic and thermo-inhibitory centers have their functions so nicely balanced in the normal adult nervous mechanism that, with the aid of the heat-dissipating centers, they are able to maintain the body at almost an uniform temperature under the most adverse circumstances, and this temperature equilibrium can only be disturbed by some maladjustment of this nervous mechanism, which would produce either increase or decrease of the body temperature.

---

* Metabolism centers might be a better name for these centers.

**Where are these heat centers located?** Ott,
Richet, Sachs, Aronson, Wood, Reichert, Girard, Ba-
ginski and White agree that the dominating thermogenic or heat producing centers are situated at the
base of the brain in or near the corpus striatum.
Eulenberg, Landois, Wood, Ott, Reichert and
White agree that important thermo-inhibitory centers are located in the cerebral cortex, and they are
known as the cruciate and Sylvian centers.

As a prelude to the use of these physiologic data
in the explanation of some important clinical phenomena associated with the diseases of infancy and
childhood, let us first inquire, what should one expect, in the light of these facts, would be the influence on the body temperature of disease or injury
of the parts of the brain containing these centers?

1st. *Destruction* of that portion of the cerebral cortex containing the cruciate or Sylvian inhibitory
heat centers should cause a rise of temperature because the inhibitory influence of these centers on
the basal thermogenic centers would be wholly or
partially withdrawn. Experimental physiology
confirms this deduction. This is probably the explanation of the fever that follows cerebral hæmorrhage into the cortex, and a partial explanation
of the fever of insolation.

2d. *Irritation* of these cortical inhibitory centers should cause a sub-normal temperature by strengthening the inhibitory control which they exercise over the thermogenic centers ; this is also evidenced by physiologic experiments. We have here an explanation of the sub-normal temperature which may result from cortical meningitis and from hæmorrhage, foreign bodies or depressed bone, all of which may first act by irritating these cortical centers (sub-normal temperature), and later by destroying them (increase of body temperature).

3d. *Destruction* of the basal thermogenic centers should cause a decrease of the body heat. But clinically there is little opportunity to observe the effect of destructive lesions of this portion of the brain, since any lesion sufficiently severe to destroy the basal heat centers would cause immediate death by the involvment of adjacent centers controlling vital processes. In shock we possibly have an example of sub-normal temperature from partial paralysis of these centers, and in the compression stage of basilar meningitis we may have a sub-normal temperature due to enfeeblement of these centers.

4th. *Irritation* of the basal thermogenic centers should cause an increase of body heat, this fact which is proven by physiologic experiment is the explanation of the increased temperature that accompanies the specific fevers.

When are the heat centers developed? The answer to this question is in great part the answer to the question why are infants and children more prone to high temperatures than adults? The heat dissipating centers situated in the medulla oblongata are well developed at birth, but these centers because of their special clinical importance in infancy and childhood will be given separate consideration later on. Here it is my purpose to note and especially emphasize the time of functional development of the heat-producing and the heat-inhibiting centers.

Before birth the thermogenic centers are in a state of immature functional development. In the human infant born prematurely they are so imperfect that artificial heat is necessary for a time to keep the body heat up to the normal. In this respect the immature human fœtus resembles cold blooded animals who are more or less dependent on their surroundings for their body heat. But as the fœtus matures, the thermogenic mechanism reaches a state of fair development, so much so, that one may say that the thermogenic centers are functionally competent at birth; this of course must

be so, since the formation of body heat is a vital
process, and is as we have seen probably controlled
by the same mechanism that controls the all important process of tissue metabolism.   While the
thermogenic heat centers have a fair degree of development at birth, they are yet immature and unstable, and are therefore like all the nerve centers
in the unfinished brain of the child, more easily excited to abnormal action, than are the mature heat
centers of the adult brain.   All the nerve cells of
the rapidly growing brain of the infant and child
are in a state of more or less structural instability
since the metabolism going on within them must
not only be rapid enough to supply waste but also
to furnish material for the growth and development
of new cells.   This structural and functional instability of the cells makes them more irritable and excitable than the nerve cells in the finished brain of
the adult.   For this reason one would expect to
find the thermogenic heat centers of the child more
excitable than those of the adult, and such in fact
is the case.   This is one important reason why the
temperature of the infant is so variable and unstable under slight disturbing influences and why like
causes produce higher temperatures in the infant
and child than in the adult.

But important as this normal excitability of the immature thermogenic centers of the child may be, yet of far greater importance, from a clinical standpoint, is the greatly increased irritability from unfavorable conditions of heredity, nutrition and environment. The thermogenic heat centers of the nervous, anemic, delicate child are in a state of abnormal excitability, so that a slight excitation may produce an abnormal discharge of force resulting in fever and high temperature.

But after all probably the most important cause of the instability of temperature in infauts and children is to be found in their feeble cortical thermo-inhibitory centers. The thermo-inhibitory centers, like other cortical inhibitory centers previously spoken of, have very imperfect functional development at birth, so that at this time they do not exert a very strong controlling influence over the basic thermogenic centers, and are not able to inhibit these centers from discharging increased energy under slightly increased excitation ; for this reason slight causes may produce an elevation of temperature in the infant. Hale White says in speaking of the thermo-inhibitory centers : " In the human adult they are fairly competent and active as is proven by our pretty constant temperature."

"In the lower animals and in children they are probably not so completely evolved for I have found that the normal temperatures of rabbits varies several degrees, and rapid fluctuations of temperature are common with children when slightly ill."

Ott in a recent personal communication says : "It seems to me that children are more prone to high temperatures because of a loss of control of the cortical centers."

It is on the whole a justifiable conclusion from all the evidence in our possession that the high and variable temperatures of infancy and childhood are in part due to the normal immaturity and instability of the cortical thermo-inhibitory centers. But as I have previously noted the feeble inhibition in the normal child is not of so much clinical importance as the abnormally feeble inhibition of the abnormal child, this is as true of the heat regulating mechanism as it is of all other nervous mechanisms. The inhibitory part of the heat mechanism in its feeble and unstable state is the portion of this mechanism which suffers most from disease, and in its development is still further retarded by unfavorable conditions of heredity and environment. McAlister says : " The inhibitory is the first portion of the heat regulating mechanism

to fail under injury or disease." All of this is
quite in accord with the general observation previously made that the amount of energy developed
by a nerve cell will depend directly on the amount
of healthful chemical metabolism going on within it.
The maximum amount of energy being stored up in
the well-nourished cell and the minimum amount
of energy in the starved cell. One can readily understand then how a malnutrition of the nerve elements resulting either from heredity, impoverished
blood or bad hygiene can still further weaken the
physiologically incompetent cortical thermo-inhibitory centers of the child, so as to make it more
prone to variable and to high temperatures from
slight causes than the normal child is, since in this
condition the energy from the thermogenic centers
would be discharged under much less restraint from
the inhibitory centers than it is in the normal child.
It may not be out of place here to state that the
best explanation we have for the rapidly varying unstable temperature that not infrequently occurs in
hysterical women is, that it is due to the instability
of the cortical thermo-inhibitory centers which have
given way under the combined influence of environment, bad heredity, bad hygiene and impoverished blood.

From what has been said the following summary
may be made of the reasons why children are more
prone than adults to high and variable temperatures :

1st. In normal children the thermogenic centers
are more unstable and therefore more easily excited
than in the adult.

2d. In normal children the thermo-inhibitory
centers are weaker, more excitable, and therefore
more incapable of exercising proper control over
the thermogenic centers, than they are in adults.

3d. In nervous, anemic children the thermogenic
centers are far more excitable than in the normal
child, such children are therefore more prone to
high and variable temperatures.

4th. In nervous, anemic children the thermo-inhibitory centers are even weaker than in the normal child, and therefore still more incapable of restraining the discharge of force from the thermogenic centers; this is a most important reason for
the variable and high temperatures of such children.

**Exciting causes of fever and high temperature
in infants and children.**  Having studied the peculiarities of the nervous mechanism which controls
the body temperature of the infant and child, we will

now inquire what are the causes most likely to disturb this mechanism so as to produce an increase
or decrease of body temperature. Or in other
words, we will ask what are the usual exciting
causes for the high and variable temperatures which
are so likely to occur in infancy and childhood?
These causes may be classed as follows :

1. Bacterial products.
2. Insolation.
3. Muscular action (convulsive).
4. Mechanical and reflex causes.

1st. **Bacterial products** are by far the most important of the exciting causes of fever and high temperature in children. The variations in temperature accompanying the acute infections, including
all forms of external and internal bacterial toxemias, are due to the action of bacterial products on
the heat centers. Bacterial products capable of
producing fever and variations in temperature may
be formed by bacterial action either within the
blood and tissues of the animal or outside the
blood and tissues of the animal, in wounds, or in
cavities, such as the intestinal canal, which com-
municate with the external air. But wherever
these bacterial products may be formed the soluble
ones are absorbed and produce fever and variable

temperature by their direct action on the nervous
centers. As a rule the soluble bacterial products
which produce fever also produce increase of body
temperature and the increase of temperature is
often a valuable indication as to the severity of the
fever process, but this is a rule which unfortunately
has many exceptions, as is shown by the subnormal
temperature that occasionally attends pneumonia,
malaria, typhoid fever, influenza, scarlatina, and
other acute infections. The subnormal temperature which occasionally occurs in these fevers has
not been satisfactorily explained. Very recently
Centanni investigated seventeen pathogenic species
of bacteria and found in cultures of all of these
germs a substance, not a peptone, which when injected into animals caused fever with the following
symptoms, high temperature, prostration, emacia-
tion, and finally death. Omitting further discussion
I will say that the evidence justifies the conclusion
that bacterial products excite fever by acting directly on the fever * centers and the variations in
temperature that accompany fever are due to the
action of bacterial poisons on the heat * centers.

---

*The fever and heat centers are probably identical since ex-
perimental physiology has not been able to differentiate between them.

Why do bacterial products produce fever and variable temperature so much more readily in children than they do in adults? This question has in part been answered by our previous study of the pecularities of the heat mechanism in childhood.

(a) **The** thermogenic centers being more unstable and irritable in the child are more readily excited by bacterial products. Fever and increased temperature are therefore more easily produced.

(b) The thermo-inhibitory centers being immature and feeble in the child they exercise but a weak restraining influence over the discharge of force from the thermogenic centers which are being excited by bacterial products. For this reason fever and increased temperature are more easily produced by bacterial products in the child than in the adult.

(c) Still another possible reason why microbic poisons produce fever and increase of temperature more readily in the child than in the adult was suggested to me by Prof. Charles Richet in a personal communication. This explanation depends on the potency of the fever poison and not upon the peculiarities of the nervous mechanism. Richet asks : "Is it not possible that the microbic fever producing toxins may be stronger or more toxic when they are produced in young organisms that are not protected by previous attacks of acute infections?" That is to suggest that in infants and children who have not had previous microbic infection and who are not therefore protected against these diseases pathogenic microbes may develop more potent fever producing toxins than they can later in life.

2d. **Insolation** is an important cause of fever and high temperature in infancy and childhood. The best explanation of the fever of insolation is that the feeble inhibitory heat centers of the child are still further weakened by the heat so that practically no restraint is exercised over the heat producing centers. Cases of insolation in infancy and childhood are ordinarily classed as cholera infantum, or other forms of summer complaint, and this classification greatly obscures the direct etiological importance of heat in these cases. Forchheimer has for many years taught that many of the cases of so-called cholera infantum were cases of insolation, and that in such cases the intestinal fermentation is primarily a symptom rather than a cause of the disease.

3d. **Convulsive muscular action** is not an infrequent cause of increased temperature in infants and children. The manifestation of muscular energy is always accompanied by the evolution of heat and the production of carbonic acid, and excessive muscular action such as occurs in general convulsions is always accompanied by increased production of heat. And this is a factor of the increased temperature that occurs in general spasms. But a portion of the increased body heat that occurs in this condition may be attributed to the increased friction of the muscles, tendons and articular surfaces which transform kinetic energy into heat. It should be remembered therefore that excessive muscular action may be a factor in producing increase of body heat and that this source of heat production is quite distinct from that which results from the normal metabolism constantly going on in the muscles, etc., at rest, and from the abnormal metabolism going on in the muscles, etc., during fever. I do not wish to convey the idea that increased muscular action is the most common or most important cause of the increased body temperature that occurs during muscular spasm, but only to impress the fact that violent muscular action is a factor in producing the increased body heat rather than that the increased body heat is a factor in producing the spasm.

When the spasm is purely reflex in origin the excessive muscular action is then no doubt the most important cause of the increased body heat, but when the spasm results from microbic poisons, as it usually does, then no doubt the increase of temperature is chiefly due to the action of these poisons on the heat centers. For these reasons one would expect to find the temperature during reflex spasm not so high as it is in spasm due to microbic infection.

4th. **Mechanical and reflex causes** of fevers and the variable temperatures of infancy and childhood. In speaking of the heat centers I have already indicated how foreign bodies, growths, and exudations could act directly on the heat centers to disturb the body temperature, so that there now only remains the consideration of the reflex causes of variations in the body temperature of infants and children. Ott says: "After the use of large doses of atropine I have seen the temperature rise greatly upon sciatic irritation. * * It was also found that this increase of temperature was accompanied by an increased production and augmented dissipation of heat." In these experiments we have proof that not only high temperature but also fever may be produced reflexly.

**Gall stone fever,** which is classed by physiologists as a reflex fever, has been studied by Wood, who found that in the "fever produced by gall stones, elevation of temperature did increase urea elimination." Here again is an instance of both fever and increased temperature from a purely reflex cause. It is my belief that variations in the body temperature in infancy and childhood are not infrequently of reflex origin, and that the intestinal canal and the genitalia are the sites where reflex irritation is most likely to produce this symptom. Increased temperature may occur in the infant and child from the cutting of a tooth, from worms, undigested food, and other foreign bodies in the intestinal canal. The irritating products of an intestinal fermentation may also produce increase of temperature unaided by the soluble bacterial poisons previously spoken of. It is a matter of every day experience with clinicians that the removal of such simple causes as are here narrated will ofttimes cause the temperature of the sick child to fall to normal and all the other symptoms of fever to disappear. It will be well therefore, in these days when chemistry and bacteriology are dominating medical pathology, for us to remember that a purely reflex fever can and does sometimes occur during infancy and childhood.

# IV.

## HEAT DISSIPATING MECHANISM.

In the previous chapter the consideration of the heat dissipating mechanisms was purposely omitted because it was thought that certain clinical phenomena dependent on the peculiarities of this mechanism during infancy and childhood could best be studied in a separate chapter.

The heat dissipating mechanism is the mechanism by which we keep ourselves cool.  This may be done in three ways :

(1) By radiation and conduction of heat from the surface of body.

(2) By constant evaporation of water from the surface of body.

(3) By evaporation of water from the air passages.

Dissipation of heat by radiation from the surface of the body is by far the most important means of heat dissipation.  In this process the vaso-motor nervous mechanism is all important.  When un-

usual heat loss is demanded the vaso-motor nerves
dilate the blood-vessels of the skin and in this way
expose more blood to the lower temperature of the
air.

Loss of heat by evaporation is dependent on the
activity of the sweat glands which are controled
by sudoriparous nerves and sweat centers. When
unusual heat loss is demanded these centers respond
by increasing the activity of the sweat centers
which cover the surface of the body with fluid and
the temperature is lowered by its evaporation.
Both the dominating vaso-motor and sweat centers
are located in the medulla oblongata and have
reached good functional development at birth.
But in the infant and child they respond more
readily and energetically to the demands for heat
reduction than they do in the adult.

It must also be kept in mind that heat loss from
both radiation and evaporation is greater in the
infant than in the adult because its area of surface is greater in proportion to its body weight,
the infant has in fact a three-fold greater radiation.
These are the reasons therefore why the high temperatures of infancy and childhood are so readily
reduced by the heat dissipating mechanisms. The
increased activity of the heat dissipating mechan-

ism acting on a proportionately larger surface
compensates for the increased activity of the thermogenic centers.  In the play of function between
the heat generating centers and the heat dissipating
centers we have an explanation of the rapid variations of temperature so characteristic of the fevers
of infancy and childhood.

**Evaporation of water from the air passages**
is a means of heat dissipation which we have yet to
consider.  And it is the special purpose of this
chapter to study this function in its relation to
clinical phenomena.

In certain animals, the dog for instance, who do
not sweat, the evaporation of water from the air
passages is the chief means of reducing the body
temperature.  Richet calls the rapid respirations of
the panting dog *Polypnœa.*  By these rapid respirations, amounting to as many as 400 in a minute,
the heat of the body is rapidly given off.  Richet
located the polypnœic center in the medulla oblongata.  Ott later located it in the tuber cinerium.
Richet proved that the polypnœic center was not
affected by the amount of carbonic acid or oxygen
in the blood, and that it was solely for the purpose
of heat dissipation.

In answer to the question, How is the polypnœic

center excited to activity? we have the experiments of Sihler demonstrating that increased respirations of an animal exposed to heat is due to two causes, warmed blood and stimulation of the skin by the heat, and that skin stimulation is the more important factor. Gad and Mertschinsky also demonstrated that increased temperature of the blood stimulates the respiratory centers and causes an increased number of respirations, and Ott produced polypnœa by electrical stimulation of the tuber cinerium.

Does the polypnœic center exist, and is it functionally active in infancy and childhood? The answer to this question has most important clinical bearings. Ott says, " In infants we see a polypnœa during fever, the respirations rise in frequency with the rise in temperature." Every physician must have seen many cases of rapid respiration in children that could not be accounted for by pulmonary disease. It not infrequently happens that a child with fever will have 60, 80 and 100 respirations per minute, without presenting any sign or symptom of lung trouble. Polypnœa is to my mind the only explanation of this phenomenon. Very rapid breathing is a common symptom of summer complaint, and in many cases means nothing more than nature's attempts at heat dissipation. The importance of recognizing polypnœa as a symptom of fever in infancy and childhood, is very great. If we do not do this we may often be led, by the rapid breathing, away from the real cause of the disease. Fortunately for us as clinicians there is a marked difference between the character of the polypnœic breathing and the rapid respirations due to lung or heart disease. In polypnœa the breathing is regular, easy and rapid, but is not as it is in lung and heart disease irregular, labored and accompanied by cyanosis.

# V.

## AUTOGENETIC AND BACTERIAL TOXINES.

The qualitative and quantitative changes in blood supplying nerve tissues are, from a clinical standpoint, the most important causes of the neuroses of childhood.

The importance of blood changes as a cause of neurotic disease depends, not only upon the fact that they are most potent factors, but more because they are factors which can as a rule be removed by treatment, and as clinicians we are especially interested in the remediable causes of disease.

For convenience of study, one may say that there are four important blood changes related to neurotic disease:

1st. The presence of autogenetic toxines in the blood.

2d. The presence of bacterial toxines in the blood.

3d. A venous condition of the blood.

4th. An impoverished condition of the blood.

The above named blood changes do not, as a rule, exist as separate pathological conditions; but should rather be considered as factors of a complex blood condition which is very commonly etiologically related to the neuroses of childhood. The above classification is given that it may furnish topics for discussion.

**Auto-intoxication.** Auto-intoxication is one of the most important and certainly one of the least understood of all the causes of neurotic disease both in adults and children. The poisons of this class are not of microbic origin, but they are for the most part either substances which are formed by the various organs of the body to serve some physiological purpose, but which are toxic when abnormally accumulated within the blood and tissues, or they are substances which are either normally or abnormally formed in the tissue changes incident to the functional activity of muscles and other organs. The poisons which produce auto-intoxication are therefore as a rule substances which are normally produced in the body in such quantities that they can readily be disposed of by the tissues or be eliminated by the intestinal canal, the kidneys, liver, lungs and other excretory organs. In this way these bodies may be excreted as rapidly as they

are formed, so that under normal conditions they
do not accumulate in sufficient quantities to produce nervous or other symptoms. But in certain
pathological conditions there may be such an accumulation of these poisons that they become most
important factors in the production of disease; this
may result either from a perverted metabolism, which
causes an increased production of these poisons, or
from disease of the kidneys, liver, or other excretory organs, which will cause their accumulation
from defective elimination or defective neutraliza-
tion.

But vast and important as is this field of the relation of auto-intoxication to nervous disorders,
yet it is so confused and so full of misinformation
that it seems almost presumptuous to write upon it.
One begins to realize what an important disease
producer auto-intoxication is, when told that it is
the most important etiological factor of acute and
chronic uremia, of gout, migraine, migrainous
gastric neurosis, migrainous epilepsy, neurasthenia,
hypochondriasis, neuralgia, myalgia and possibly
other nervous disorders. The importance therefore
of this field, as well as the darkness which shrouds
it, are my excuses for attempting its exploration.

Bouchard, in his "Lectures on Auto-intoxication," proves that normal human urine is toxic, when injected in large quantities into the veins of rabbits; he found on the average that it requires forty-five cubic centimeters of urine to kill one kilogram of rabbit. From this Bouchard estimates that, "On an average of two days and four hours man makes a mass of urinary poison capable of intoxicating himself." Bouchard's experiments were conducted with sufficient care to prove that normal urine contains toxic bodies; but the large quantity of urine necessary to produce intoxication also proves that the poisonous bodies in *normal* urine either exist in minute quantities, or have a very low degree of toxicity. Bouchard from his researches concludes that there are seven toxic principles in normal urine.

1st. "A diuretic substance," urea, which, by reason of this property, "plays a useful rôle in the economy." While urea in "an enormous dose" may be said to be toxic, yet "there are few bodies in the urine so feebly toxic as urea, if we except albumen and water which naturally exist in the blood." "Sugar is more toxic than urea." "Urea has almost the toxicity of the most inoffensive salts." These observations concerning urea are in accord with the well established physiological fact, that urea is not sufficiently poisonous to play any part in the production of urinary toxemia.

2d. An unnamed narcotic body, which has not been separated from the urine, and which is thought to be the cause of the narcosis produced by the injection of normal urine.

3d. An unnamed sialogogic body which has not been separated from the urine. It is presumed to exist in minute quantities in human urine, because under certain conditions urine produces salivation.

4th. An alkaloidal body "endowed with the property of causing convulsions." This body has not been named or isolated.

5th. An organic substance which contracts the pupil, and causes convulsions. It has not been named or isolated, but is thought to be a coloring substance.

6th. An organic substance which "reduces heat," not named or isolated.

7th. The potash salts, "whose convulsive properties have long been known," play an important part in the production of urinary toxemia.

Bouchard believes that the above named bodies play an important rôle in uremia, and other auto-intoxications. But important and valuable as this work is, it really gives very little definite knowl-

edge except that normal urine is feebly toxic, and
that it contains a number of unnamed and uniso-
lated toxic bodies, which are so feebly toxic, or exist in such small quantities, that it requires " on an
average of two days and four hours for a man to
make a mass of urinary poison capable of intoxicating himself."

On the whole one may say that Bouchard's work
is a valuable contribution to the study of auto-intoxication, but it falls far short of giving a satisfac-
tory explanation of the symptoms of " uremia " or of
other auto-intoxications.

It is my belief that pathological, rather than normal urine, holds the most important secrets of auto-
intoxication, and I would ask attention to a phase
of this subject to which I have given much study,
viz., the relation of the uric acid diathesis to nervous diseases.   On this subject I wish to speak
emphatically, in the hope and belief that what I
shall say will throw a ray of light into one of the
darkest fields in medical pathology.   At the present time there is no fallacy so deeply rooted in the
medical mind, as that uric acid and urea can, by
their direct action on the nerve centers, produce
nervous disease.   This belief has been so firmly
fixed in the medical mind, that it was considered al-

most too trite a fact for medical discussion, and even
now it seems almost heresy to proclaim that uric acid
and urea do not produce nervous symptoms, since
the ingenious theory of Haig, that migraine and
kindred nervous disorders are commonly due to an
excess of uric acid in the blood, is now very generally accepted by the profession.

In the field of pediatrics we have blamed uric
acid with causing paroxysmal gastric neuroses, migraine, convulsive disorders, and other neurotic diseases; while against urea has been charged certain
convulsive symptoms. But I wish here to emphasize the fact, that both urea and uric acid have, in
this regard, been falsely accused, and convicted upon
circumstantial evidence. They are innocent, non-
poisonous bodies, not capable of producing the
severe nervous symptoms, which accompany the
excessive elimination of uric acid and the diminished elimination of urea.

Bouchard injected experimentally in the blood
thirty centigrams of uric acid for each kilogram of
animal without apparent injury. In one instance
he injected sixty-four centigrams for each kilogram of animal, without injury to the animal.

Roberts says: "Uric acid and its compounds are
deleterious simply because of their sparing solubil-

ity in the body media." In fact all experimental
evidence is opposed to the idea that uric acid or its
compounds can, in any other than a mechanical way,
produce nervous symptoms. Uric acid has been
accused and convicted of producing certain nervous
disorders, on the circumstantial evidence that it was
present in excess in the urine immediately before,
during, or after an attack of these diseases. But
its innocence is now thoroughly well established,
and the same may also be said of urea.

In the light of our present knowledge, we can
only consider the increase of uric acid and the diminution of urea in the urine as valuable signs, in-
dicating the approach, or presence of a nervous
attack, due, in all probability, to auto-intoxication,
but of which they are entirely innocent. Sir Alfred
Garrod, Sir Dyce Duckworth, M. Lecorche, Sir
William Roberts, Murchison, Alexander Haig and
many other observers have conclusively shown that
there is a definite relation between the quantity of
uric acid excreted and the paroxysms of gout; and
several of these observers have also called attention to
the relation existing between the amount of uric
acid and urea excreted and certain paroxysmal nervous diseases, such as migraine, epilepsy, spasmodic
asthma and uremic manifestations. But no one

has given so much time and careful study to the relation that exists between nervous diseases and the presence of urea and uric acid in the blood as has Alexander Haig.

The term uric acid diathesis is, therefore, rightly used to describe the condition in which, either from increased production, or deficient elimination, there is an excess of uric acid and its compounds either in the blood or in the tissues, which are closely associated with gout, migraine and allied diseases. In some disorders, such for example, as gout and gravel, which belong to the uric acid diathesis, the distressing symptoms are, no doubt, due in part to the precipitation in the tissues, or elsewhere, of the comparatively insoluble urates; but in other diseases, such for example, as migraine which may also under the above definition be classed as coming under the uric acid diathesis; the uric acid, although it may occur in excess in the blood or urine in these cases, has nothing whatever to do with the production of nervous symptoms. In accepting the non-toxicity of uric acid and its compounds, one must not forget that, by reason of their insolubility, these substances may be important pathological factors in gravel and articular gout, and that they may, in a reflex way, even produce nervous

symptoms; and it should also be remembered that,
by their presence in excess in the urine, they may
serve as important signs or signals to announce the
presence or the approach of migraine and allied
nervous diseases.

In the light of the above facts we are left absolutely without any explanation of the constitutional
symptoms which occur in the uric acid diathesis,
and which have heretofore been ascribed to the
toxic influence of this body on the nervous centers. In this condition of affairs, one may therefore be pardoned for suggesting, that the poisonous
bodies so closely allied to uric acid, and named
and classified as leukomains, may be in part responsible for the nervous symptoms that have heretofore been attributed to uric acid. For it seems
altogether possible that, along with the increased
excretion of uric acid, there might also be an increased excretion of uric acid leukomains, since
these bodies belong to the same chemical group,
and are probably formed by the same, or a like
metabolism. It also seems possible that a perverted
metabolism, or a defective elimination, might result
in these leukomains being present in the blood in
such abnormal quantities as to make them in
part responsible for the nervous symptoms. Some

of these leukomains, notably paraxanthin, gerontin, and xanthin, are very poisonous, and quite capable of producing nervous symptoms if they occur even in small quantities in the blood. The uric acid leukomains are a group of bodies closely related to uric acid, of which paraxanthin, xanthin, and gerontin are poisonous, and xantho-creatinin is a poisonous leukomain of the creatinin group. It is quite possible that all of these poisonous leukomains, as well as other unnamed poisonous leukomains, may contribute to the production of the complex of symptoms which I shall here class under leukomain poisoning; but in this study I shall only attempt to show that paraxanthin and xanthin are etiologically related to the group of nervous disorders above noted as being manifestations of leukomain poisoning. Paraxanthin is by far the most poisonous of all known leukomains. Salomon thus describes paraxanthin poisoning in the mouse. If one-half milligram of paraxanthin be introduced into the peritoneal cavity of a mouse the following symptoms will result: " The reflexes are increased to a tetanus, followed by a rigor-mortis-like contraction of the muscles; marked dyspnœa is a constant symptom which continues till death."

Xanthin is very much less poisonous than parax-

anthin, but according to Filehne it produces in the
frog a decided muscular rigor and paralysis of the
spinal cord.  In brief, we may note the following
facts concerning xanthin and paraxanthin as bearing on this subject :

1st. Paraxanthin and xanthin are poisonous leukomains of the uric acid group, capable of producing the most profound nervous symptoms.  They
are readily soluble in water, urine and blood.

2d. Paraxanthin is found in normal urine in
such small quantities that its poisonous properties
are lost in dilution.  Salomon found only 1.2 gm.
in 1,200 litres of urine.  This quantity is so minute
that its presence can not be satisfactorily demonstrated in such quantities of normal urine as can
conveniently be obtained from patients.  In a recent personal communication Salomon says : " Nine
litres of normal urine is a very small quantity to
prove the presence of paraxanthin if one has not
previously worked with larger quantities so as to
master the details of the work, and very much
harder would it be to prove the presence of paraxanthin in four litres of normal urine, as I know
from experience.  .  .  .  I would advise that not
less than ten litres of normal urine be used to
demonstrate the presence of paraxanthin."  My own

experience is in accord with Salomon's. In previous papers I have recorded my failure to demonstrate the presence of paraxanthin when working with as little as four litres of normal urine; and, since these papers were written, I have made a large number of examinations of normal and other urines, and I have always failed to demonstrate the presence of paraxanthin in four litres of normal urine. Upon this evidence I have concluded that paraxanthin is present in abnormally large quantities when I can find it in less than four litres of normal urine. Xanthin also, as a rule, requires more than four litres of urine to demonstrate its presence, but I have frequently found small quantities of xanthin where I could not find paraxanthin in working with four litres of urine.

3d. Paraxanthin and xanthin are not formed in the kidney. They are excreted from the blood by the kidneys. The presence, therefore, of large or small quantities of xanthin bodies in the urine means that these bodies were present in large or small quantities in solution in the blood previous to their elimination by the kidneys.

With the above facts in mind concerning xanthin and paraxanthin, we are better prepared to study leukomain poisoning, which, I believe, is the most important form of auto-intoxication.

**Leukomain Poisoning.** In the Medical News, Philadelphia, May 26, 1894, I published a paper on "Paraxanthin as a Factor in the Etiology of Certain Obscure Nervous Conditions." That paper was based on the study of a patient who had migraine all her life till she was past sixty years of age, at which time the migrainous attacks were superseded by epileptoid paroxysms, which came at about the same interval of time as the migrainous attacks had previously come. These epileptoid attacks were very severe, and very sudden in their onset; almost immediately the muscles would become rigid and the breathing would be labored, gasping and irregular; the heart's action was rapid but would remain regular and strong. These attacks would last from twenty minutes to an hour and unconsciousness would continue from the beginning to the end. In the interval between these attacks the patient was well mentally and physically. By a careful study of this case, both before and after (See Medical News, Philadelphia, November 3, 1894, and Medical Record, New York, June 22, 1895) the publication above referred to, I have demonstrated that the epileptoid symptoms of this patient were caused by the direct action of poisonous uric acid leukomains on the nerve centers. I found that the urine of this patient passed during and just after an attack of epilepsy contained an excess of uric acid, and that in color, quantity and other particulars it corresponded to the urine of the epileptoid cases which Alexander Haig and others have thought to be due to uric acid ; but I also found, a fact of much greater importance, and one that has previously been overlooked in the study of these cases, viz., that along with the excess of non-poisonous uric acid compounds excreted during and after these attacks, there was also excreted in the urine *a very great excess of paraxanthin and other poisonous uric acid leukomains.*

The paraxanthin solution obtained from the urine of this patient, when injected into rats and mice, produced epileptoid symptoms very similar to those from which my patient suffered when this same paraxanthin was circulating in her blood just prior to its excretion by the kidneys.

From the study of this and other so-called migrainous epilepsy cases I am convinced that there is a form of epilepsy which begins as a rule in middle life, either alternating with or taking the place of migrainous attacks of previous years, which has as its most important etiological factor the presence in the blood of the very poisonous leukomain paraxanthin. This form of epilepsy is an auto-intoxication—a true leukomain poisoning. A further study of these cases leads me to believe that many of the hystero-epilepsy cases are leukomain epilepsies, and are therefore more amenable to medical than to surgical treatment.

It is impossible at the present time to say just what part paraxanthin poisoning plays in the production of puerperal and other eclampsias of uremic origin, but it is not improbable, in fact it is my belief, that the poisonous leukomains are in part responsible for uremic symptoms.

**True migraine** is perhaps the most common of the well marked forms of auto-intoxication due to leukomain poisoning. In the Medical News, Philadelphia, November 3, 1894, and the Medical Record, New York, June 22, 1895, I have published a study of a number of cases of true migraine, in which I demonstrated that attacks of migrainous headache were always immediately followed by the excretion of an excess of uric acid in the urine of these patients; this fact has been fully worked out by Alexander Haig and other English writers who have unjustly accused the non-poisonous uric acid of producing the complex of nervous symptoms so characteristic of migrainous attacks. But these and all other investigators have up to the present time overlooked the fact demonstrated by me in the papers above referred to, viz., that along with an excess of uric acid, there is excreted, in the urine passed just after an attack of migraine, a very great excess of xanthin and paraxanthin, and that the solution of these leukomains, obtained from such patients, produces in rats and mice the characteristic symptoms of paraxanthin poisoning.

From my researches it is plain that attacks of migrainous headache are coincident with an excess in the blood of uric acid leukomains. The conclusion therefore seems justifiable, that migraine is a manifestation of leukomain poisoning, and is not, as Haig and others have thought, due to uric acid and its compounds which are also present in excess in the blood during these attacks.

There is a **leukomain gastric neurosis,** the study of which has been of more interest and more value to me than any other phase of leukomain poisoning. In the Medical Record, New York, June 22, 1895, I have published a careful study of one of these cases, in which I demonstrated that the gastric attacks were followed by the excretion in the urine

of an excess of uric acid and its compounds, and of
a great excess of xanthin and paraxanthin; and in
this case I also demonstrated the presence of
xanthin in the mucus which was discharged in great
quantities from the stomach during these attacks.

In leukomain gastric neurosis there is almost always a personal history of migraine.  The gastric
attacks often take the place of migrainous head-
aches, or the patient may suffer from both of these
manifestations of leukomain poisoning at the same
time.

The day before a gastric attack the patient may
be uncomfortable, with slight pain in the stomach,
and eructation of gas, and the urine may be very
scant and high colored.  These warning symptoms may be more or less distinct, and then the attack bursts with great fury.  At once the patient
has great pain in the stomach, and vomiting comes
on at the same time.  The pain and vomiting continue and a large quantity of glairy mucus is discharged from the stomach.  In severe cases, eructation of gas, pain in the stomach, and vomiting of
mucus continue in paroxysms till they are relieved
by the hypodermic injections of morphine.  The
frequency of these attacks and the relief obtained
from morphine gradually induces the opium habit

in the unfortunate victims of this phase of leukomain poisoning. It is therefore of the very greatest importance that physicians every-where should
recognize that these gastric attacks are of leukomain
origin, in order that they may be relieved by proper
treatment and saved from the morphine habit.
Fortunately many of the cases of leukomain gastric
neurosis are much less severe than the type of cases
just described. In some instances the attack is
terminated by the first paroxysm of vomiting, and
in others there may be no vomiting at all. In the
cases where there is no vomiting patients often
complain of periodic diarrhea, with more or less
constant pain in the stomach, and with well marked
symptoms of hypochondriasis or neurasthenia.

It is my belief that the importance of leukomain
poisoning as a disease producer is not half told by
the above outline, but it would not be profitable to
attempt to predict the findings which may come
from this field of work. In this connection, however, I shall say that· I have unpublished experimental evidence that causes me to believe that the
gastric attacks in lead poisoning, and the arthritic
paroxysms of true gout are due to leukomains.

**Biliary toxemia** is a form of auto-intoxication resulting from the absorption of bile. Bouchard has

shown that the biliary salts and the biliary coloring matters are active poisons, the latter being much the more poisonous. These substances when injected into the veins of rabbits are very active poisons, killing in convulsions. From his experiments on rabbits, Bouchard estimates that man forms in eight hours enough biliary poison to kill himself. But these experiments are not fully substantiated by clinical experience, since the absorption of considerable quantities of bile may go on over a long period of time, producing a well marked jaundice, without causing very pronounced symptoms of auto-intoxication. The simple jaundice of infants and the catarrhal jaundice of children do not, as a rule, have well defined symptoms which can be ascribed to the toxic influence of bile. But children suffering from the more severe forms of jaundice may be irritable, and may even have convulsions; or they may be drowsy, stupid and pass into coma; but it is impossible to say what part the biliary toxemia plays in producing these symptoms. In the present state of our knowledge, the further discussion of this subject would not be profitable.

**Bacterial toxines.** Bacterial toxines play an important rôle in the etiology of the neuroses of childhood. We know from both laboratory and

clinical observations that bacterial products can, by
their direct action on nerve elements, produce most
profound nervous symptoms. In a previous chapter we have seen that bacterial products are by far
the most important of the exciting causes of fever
and high temperature. These bacterial poisons
produce this effect by their direct action on the
heat centers. High temperature is therefore a
nervous symptom which is, for the most part, produced by the physiological action of certain bacterial products on the heat centers. Centanni has
isolated this fever producing toxine from pure cultures of a large number of pathogenic bacteria. The
toxines produced by the tetanus bacillus were
shown by Brieger to be the cause of the profound
nervous symptoms of that disease. From pure
cultures of the tetanus bacillus he isolated bacterial
products capable of producing tonic and clonic
muscular spasm. Poisonous bacterial products capable of producing marked nervous symptoms
have also been isolated from pure cultures of a
number of other bacteria, including those of diphtheria, cholera, tuberculosis, typhoid fever, septi-
cæmia and other acute infections, so that clinicians
have now very generally come to believe that the
nervous symptoms of the acute microbic diseases

are, in great part, due to the action of bacterial toxines on the nervous system. But these acute bacterial toxemias do not properly come within the scope of my subject, and we are therefore not so much interested in them, as we are in those chronic blood intoxications which result from such chronic microbic diseases, as tuberculosis, malaria, summer complaint, rheumatism and syphilis, since these are the chronic diseases, which are so intimately associated with the neuroses of childhood.

Hysteria, incontinence of urine, night terrors, chorea and other neuroses, are very commonly associated with tuberculosis in childhood. In the Archives of Pediatrics, May, 1893, under the heading "Tuberculous Neuroses of Childhood," I noted the fact that, in my clinic, chronic glandular tuberculosis was, of all diseases, the most intimately associated with the neuroses of childhood. How then is tuberculosis etiologically related to these neuroses? We know that the tubercle bacillus produces a toxine which, when introduced into the body, causes fever and other well marked nervous symptoms; the inference therefore is unavoidable that this tuberculous toxine is a factor of that complex blood condition which results from chronic tuberculosis and which is etiologically related to many of the neuroses of childhood.

Chronic malaria is etiologically closely related to the neuroses of childhood, and this relationship is no doubt, in part, due to the production of nervous symptoms by malarial toxines. It has been demonstrated that toxic substances are produced in malarial disease, and that these substances are eliminated in large quantities in the urine during a malarial paroxysm; and what is of more importance to us in our present study is that it has been proven that the urine in chronic malaria is at all times more toxic than normal urine. These malarial toxines, eliminated in the urine, when injected into the veins of animals, will produce well marked nervous symptoms. These experiments corroborate the well grounded clinical belief that the many neuroses of childhood, so commonly associated with chronic malarial diseases, are, at least in part, due to the direct action of malarial toxines on the nervous tissues; and they also afford an explanation for the fact that such neuroses as neuralgia, night sweats, hysteria, incontinence of urine, night terrors, and chorea are occasionally cured by the specific treatment for malaria, viz., quinine and arsenic.

**Intestinal fermentation** is very frequently etiologically related to the neuroses of childhood, and
there can be no doubt but that intestinal toxemia
is one of the links which unite these two conditions.  There is perhaps no fact better established
by experimental and clinical medicine, than that
very active poisons, capable of producing the most
violent nervous symptoms, can be produced by a
putrid fermentation of albuminous material either
within or without the intestinal canal.  Booker,
Vaughn, and others have demonstrated that intestinal toxemia, producing the most violent nervous
symptoms, can result from bacterial products
formed in, and absorbed from the intestinal canal.
These acute intestinal intoxications are factors in
producing the greater number of the convulsions
which occur in infancy and the frequent rise of
temperature so common at this period of life.
These are facts known and taught every-where,
and therefore need not here be enlarged upon.

But in a study of the etiology of the neuroses
of childhood it is of especial importance to call attention to *chronic intestinal toxemia* as one of the
most important factors in producing these nervous
conditions.  It is my belief that the continued absorption from day to day of bacterial and other

toxines from the intestinal canal is a very important and much underestimated cause of neurotic disease in children. Chronic intestinal toxemia in childhood is, as in the adult, very commonly associated with both constipation and diarrhœa. If this fact be kept in mind we shall often be able to cure hysteria, night terrors, neuralgia, headache, neurasthenia and the convulsive neuroses by diet, cathartics, and intestinal antiseptics. Stomachal toxemia may also be a factor in producing neurotic disease in childhood.

An important fact not to be overlooked in this connection is that the toxemia is not the only change in the blood, resulting from chronic intestinal fermentation, which is etiologically related to neurotic disease. Forchheimer has shown that intestinal fermentation is a hemoglobin destroyer and an anemia producer; this would result in a general malnutrition which, as we shall see in the next chapter, is closely related to neurotic disease.

# VI.

## VENOUS CONDITION OF THE BLOOD.

It is a striking fact that the nervous symptoms resulting from a venous condition of the blood, supplying nervous centers, are quite the same as the symptoms produced by an arterial anemia of the same centers, due to a complete or partial closure of the arteries supplying these centers. The reasons for this are plain, since following the ligation of arteries we have not only an arterial anemia of the nerve centers, but also a compensatory venous congestion, so that in both artificial venous congestion and arterial anemia we have the nerve centers bathed in venous blood.

It is thought by Landois and Sterling "that the stimulation of the nerve centers which results from the ligation of arteries, is due to the sudden interruption of the normal exchanges of gases between blood and tissues." But it must be remembered that a venous condition of the blood means not only a decrease of O, and increase of $CO_2$, but it

also means more urea, more uric acid and more of
all the effete products of retrograde tissue metamorphosis. That is to say, that the nerve tissues
are not only deprived in part of all those substances
which are necessary for their nutrition and healthful action, but they are also exposed to the irritating
and poisonous influence of the effete products previously noted. It seems therefore a safer explanation of the symptoms, which result from experimental arterial anemia or venous congestion of
nerve centers, to say that they are caused not only
by an interruption in the normal exchange of all
substances, necessary to the nutrition and healthful
action of nerve tissues, but also by the presence in
the blood of $CO_2$, and other effete and poisonous
products.

In this connection we may note the following
physiological facts concerning the influence of the
above-named blood conditions on important nerve
centers.

A venous condition of the blood in the medulla
oblongata will stimulate the vaso-motor centers and
cause constriction of the small arteries; this has
been thought to be due to the direct stimulation of
the centers by $CO_2$ (Landois and Sterling). The

same result may also be produced by arterial anemia of these centers due to ligation of arteries.

In the medulla oblongata there is a center whose stimulation causes general spasms. This center may be excited either by a venous congestion or an arterial anemia of the medulla oblongata.

The respiratory center may also be excited by either a venous condition of the blood or by an arterial anemia.

Lauder Brunton cites the following experiment to show the relation existing between convulsive movements, and a venous condition of the blood, supplying nerve centers: "In fowls killed by Cobra poison the convulsions come on at the moment the comb becomes livid, and when artificial respiration is begun, the convulsions disappear as the comb again regains its normal color." Brunton believes this to be an instance of asphyxial convulsions, due to irritation of the higher brain centers, thus diminishing their co-ordinating or inhibiting action on the lower centers of the cord. He also says that "drugs which stimulate the circulation and increase the nutrition of the higher nerve centers, in this way strengthen their co-ordinating power and tend to prevent spasm ; alcohol and ether act in this way." That this weaken-

ing of the inhibitory power of the brain and
medulla oblongata may result from arterial anemia
as well as from venous congestion is shown by the
following experiment:

If the arteries going to the brain be ligatured so
as to paralyze the medulla oblongata, then, on
ligaturing the abdominal aorta, spasms of the lower
limbs occur, owing to the anemic stimulation of
the motor ganglia of the spinal cord (Sigm. Meyer).
That the anemic condition of the cord produced by
ligaturing the abdominal aorta is incapable of producing spasms, when the medulla oblongata is in
normal condition, is a striking example of the inhibitory influence of the oblongata centers on the
motor centers of the cord.

V. Aducco made a series of valuable experiments
on dogs. He produced anemia of the nerve centers
by cutting off a portion of the blood supply from
the spinal motor centers. He compared the excitability of these centers before and after the
artificial anemia thus produced, and in this way he
determined " the effect that partial anemia exercised on the motor centers of the cord."

Aducco concludes his paper as follows:  " The
researches I have just described have led me to
draw the following conclusions: in anemia, that is

to say when the flow of blood is diminished, the
active elements of the nerve centers are found in a
state of great excitability. In this condition, ex-
citants from the exterior act much more energetically than in the normal condition, and this
state of excitability increases, very probably, during the entire duration of the anemia. It seems to
me that one should, within certain limits, admit
that there is an inverse relation between nutrition
and the excitability of the nerve elements. This
latter augments during the time that the nutrition
diminishes."

In these conclusions, Aducco wrongly interprets
artificial arterial anemia to mean a *simple innutrition*, and concludes that the excitability of the nerve
centers is due to this innutrition rather than to the
numerous blood changes which we have previously
shown to accompany arterial anemia.

I have repeated Aducco's experiments and quite
agree with him that the excitability of the nerve
centers increases with the duration of the arterial
anemia ; but I have also shown by a series of experiments, made upon rabbits and dogs, that the
complete closure of the veins, returning the blood
from the spinal motor centers, will produce the

same symptoms that are produced by the ligature
of the arteries supplying the same spinal centers.

In these experiments, I studied the increase in
the electrical excitability in the muscles of the hind
legs as well as the increase in the reflex excitability
of these parts; and always obtained practically the
same results from ligatures of arteries as from ligatures of veins supplying the same nerve centers.

From the observations cited in this chapter, the
following inferences may be made :

1st. Both arterial anemia and venous congestion
can produce an excitable condition of the nerve
centers, and may therefore be factors in the production of nervous symptoms.

2d. The nervous symptoms resulting from arterial
anemia are very similar to those resulting from
venous congestion, and this is because in both conditions there is a venous condition of the blood supplying the nerve centers.

3d. Arterial anemia and venous congestion produce nervous symptoms by producing a *malnutrition* rather than a simple *innutrition* of the nerve
centers.

4th. Arterial anemia and venous congestion
weaken the inhibitory centers, and this results

in the discharge of force from reflex centers on comparatively slight excitation.

5th. Arterial anemia and venous congestion make more excitable both the reflex centers in the cord, and the more important reflex centers in the medulla oblongata.

The above outline will be of assistance in explaining many obscure nervous symptoms, and the following examples may be cited to indicate the important relationship existing between a venous condition of the blood and the neuroses of childhood :

1st. The venous condition of the blood, resulting from a weak or crippled heart, is at least a partial explanation of the relationship which exists between this condition of the heart and certain neuroses, such as chorea, hysteria and general nervous irritability.

2d. Rheumatism, scarlet fever, diphtheria, and other acute diseases, which ofttimes produce a weakened condition of the heart, may in this way be indirect factors of neurotic disease; (from what has been said in the previous chapter, it is scarcely necessary here to note that these diseases may also act in another way in producing nervous symptoms).

3d. Tuberculosis, chronic intestinal catarrh, and other diseases, that produce a profound chronic anemia and resulting malnutrition of the nerve centers, may be powerful factors in producing many of the neuroses, such for example as hysteria, incontinence of urine, chorea and spasm.

# VII.

## AN IMPOVERISHED CONDITION OF THE BLOOD.

In previous chapters I have noted certain blood conditions which produce nervous symptoms by their direct irritant or poisonous action on the nervous centers; and now I wish to inquire what nervous symptoms may be produced by an impoverished condition of the blood, producing either *innutrition* or *malnutrition* of nerve elements. It is most important in this study that one should always keep in mind that innutrition and malnutrition represent very different types of nutritive disturbance. By innutrition of nerve elements is meant a simple starvation of nerve elements, such as would result from a simple quantitative reduction in all the nutritive elements of the blood, which are necessary to the development and healthful action of nerve tissue.

By malnutrition of nerve elements is meant a bad nutrition or a qualitative change in the blood, such as a diminished amount of fat, of albumen, of calcium or of some other important constituent of the blood.

**Innutrition of nerve elements,** such as may result from a simple quantitative reduction of all the nutritive elements of the blood, rarely, if ever, exists as an unaided cause of disease, except possibly such an uncomplicated condition may be produced by actual complete starvation or by repeated hemorrhages. But while a simple innutrition of nerve elements may very rarely exist as an uncomplicated factor of disease, yet nerve innutrition in some more or less modified form is a constant accompaniment of all the blood conditions, which poison, irritate, or mal-nourish nerve elements ; and medical writers have always thought that the innutrition of nerve elements was in great part responsible for the nervous symptoms which are associated with all these abnormal blood conditions. For these reasons, it is most important that one should inquire into the exact rôle played by a simple innutrition of nerve elements in the production of nervous symptoms. After a careful study of this question, and numerous consultations with physiologists, it was decided that the best method of studying this question experimentally in animals, was by subjecting them to repeated bleedings or by starvation, or by both methods combined. Following out this idea, a number of rabbits were starved and bled into a condition of profound innutrition. In these experiments the rabbits were given all the water they would take, and as little food as possible consistent with life. These rabbits were kept for two or three weeks almost at the point of complete starvation, before they finally starved to death. It was found that the innutrition of nerve elements, which must necessarily have resulted from this starvation, had very little influence in producing nervous symptoms. After two or three weeks of starvation, the spinal reflexes were not noticeably exaggerated, and the electric excitability of the muscles was actually diminished. It would seem, therefore, from these experiments that a simple innutrition of nerve centers, when not assisted by other factors of neurotic disease, has little influence in producing nervous symptoms; and, it is moreover, here worthy of note, that this conclusion, drawn from physiologic experiments, is in keeping with the physiologic law noted in chapter first of this series of papers; viz., "other conditions being the same, the amount of energy developed by a nerve cell, will depend directly on the amount of healthful chemical metabolism going on within the cell.

The maximum amount of energy will be found stored up in the well nourished cell, and the minimum amount of energy in the starved cell." For the above reasons, I am led to believe that a simple innutrition of nerve centers leads to such a diminution of stored up energy in the cells of these centers, that any increase of irritability which may result from the simple innutrition of these cells, is of little moment in the production of reflex neuroses. The diminished amount of stored up energy in starved nerve cells will offset the influence of their increased irritability in the production of nervous disorders. But whether or not this is the true explanation of the observed phenomena, the fact remains, that experimental innutrition of nerve centers does not increase the reflex phenomena presided over by these centers.

Clinical medicine also furnishes evidence that a simple innutrition of nerve elements is not an important factor of neurotic disease, since it is a fact not infrequently observed, that long and repeated hemorrhages may produce a profound innutrition without causing any pronounced nervous symptoms. It is also a notable fact, that the starvation experiments which have been made, for notoriety and pecuniary benefit, by a number of persons within recent years, did not produce any increase in their nervous irritability.

All of these facts are in evidence to prove the truth of the proposition that *innutrition of nerve elements is not an important factor of neurotic disease in children.*

The clinical importance of this physiologic proposition is great, and must not be neglected. If clinicians would keep this fact in mind, it would always suggest to them the importance of carefully inquiring into the exact blood conditions present in nervous disorders. It is "bad" blood, not "thin" blood, that is a most important factor in producing neurotic disease in children.

**Malnutrition of nerve elements,** such as may result from a diminished amount of fat, albumen, calcium, oxygen or some other important constituent of the blood, very commonly exists as a factor of neurotic disease in children. This condition, which Christopher has described as a "partial starvation" of nerve elements results in making the nerve cells much more irritable, so that they discharge their force much more readily than stable, normally nourished cells would do. Such qualitatively starved cells are yet sufficiently well

nourished to store up considerable nerve energy
to be thus fitfully discharged.

While we are reasonably certain that malnutrition in the restricted sense here used is an important cause of nervous disease in children, yet,
we have very little accurate knowledge upon this
subject. We have reason to believe that a malnutrition of nerve elements is in part responsible
for the nervous symptoms which result from the
chronic blood intoxications referred to in previous
chapters; but in these instances it is impossible to
separate the symptoms produced by the malnutrition from those produced by the toxemia. Notwithstanding the very great difficulty of studying this
subject from a clinical stand-point, yet its importance demands that we should make an attempt at
conclusions from clinical observations, even though
they may not have the force of deductions from
clean physiologic experiments.

**Chronic anemia** is a term used to express an in-
constant and very complex blood condition, which
is one of the most common causes of general malnutrition and nervous disease in infancy and childhood. The chronic anemias of infancy and childhood are due to a great variety of causes, the most
important of which are tuberculosis, rheumatism,

malaria, syphilis, intestinal diseases, improper food
and bad hygiene. The blood in chronic anemia is
weak in proteids and hemoglobin, and such a condition must necessarily produce an oxygen and proteid starvation of the nerve cells, and there can be
little doubt but that this proteid and oxygen starvation of nerve tissue is an important factor in producing the nervous symptoms of chronic anemia
in childhood. But as previously stated, chronic
anemia is a very complex blood condition, which
may comprehend not only a diminished amount of
proteids and hemoglobin, but it may also mean a
diminished quantity of fat and of inorganic salts,
or an increase of the poisonous and irritating products previously referred to; yet, these accessory
conditions are probably not so constant in chronic
anemia as the diminution in proteids and hemoglobin, and there is, therefore, good clinical grounds
for the belief that a proteid and oxygen starvation
will increase the irritability of nerve elements, and
in that way act as important factors in producing
the various neuroses of childhood. In this we
have an explanation of the well-known clinical fact
that iron, and a food rich in easily digested proteids,
will, as a rule, relieve the nervous symptoms of

chronic anemia by raising the percentage of cor-
puscles and hemoglobin.

**Fat starvation,** as a form of malnutrition, can
best be studied in rachitis, which of all diseases is
the most closely related to the neuroses of infancy.
The work of Cheadle and others clearly demon-
strates that fat starvation is one of the important
causes of rachitis; and the feeding of some easily
digested fat is now accepted as a most important
means in the cure of this disease. It must not
be understood that the blood condition in rachitis
is described by saying there is a diminution in
the amount of fat, since there is always present
more or less chronic anemia, as described in the
previous paragraph, and also possibly a diminished
quantity of calcium and phosphorus; but by far
the most important blood condition is the diminished quantity of fat, since this is a constant condition, and one that we know is etiologically related
to rachitis, and especially to its nervous symptoms.
The inference therefore is probable that fat starvation is a form of malnutrition, which may predispose to laryngismus stridulus and other local and
general convulsive neuroses, so common in rickety
babies. It must be remembered, however, that fat
starvation is not the only factor in producing the

malnutrition of rachitis, any more than oxygen and
proteid starvation are the only causes of malnutrition in "chronic anemia" from other causes. Just
the part that calcium starvation plays in the etiology of rachitis, is a question in sharp dispute, and
one that can not here be discussed.

**Calcium starvation** may be studied to some ad-
vantage from the very careful experiments of W.
H. Howell, who demonstrated that the normal irritability of nerve and muscle tissue is in great part
dependent upon the proper supply of calcium to
these tissues. If the heart be deprived of calcium
salts, by feeding it with blood deprived of its calcium salts, it stops beating very soon, and this
action is so rapid that it could only result from
nervous influence. The most plausible explanation
of this fact is that the nerve ganglia of the heart,
in the absence of calcium, fail to discharge the
nerve force which stimulates the heart muscle to
contraction. If, on the other hand, the heart be
fed with a calcium solution in distilled water, it
will continue to beat for a long time. In this instance, the calcium keeps up the irritability of the
cardiac ganglia, so that they continue to discharge
nerve force into the cardiac muscle, and the heart's
action continues. In this explanation, which I

have taken the liberty to make from Howell's experiments, I have attributed to calcium an important influence over the discharge of nerve force from automatic centers; the presence of calcium in normal quantities causes these centers to discharge their nerve force into the cardiac muscle, as they normally do; and the absence of calcium inhibits the discharge of nerve force from these automatic centers, and as a result the heart stops.

If a certain amount of calcium is necessary to the normal irritability of nerve centers, and if the absence of calcium inhibits the discharge of force from nerve centers, then it is reasonable to infer that a diminished amount of calcium would have an influence on the irritability of nerve centers, which would find expression in clinical manifestations. That an insufficient quantity of calcium in the blood may produce nervous symptoms, is, I think, proven by Howell's experiments. He says: "When a frog is irrigated with oxylate solutions, that is to say calcium free solutions, the muscles are affected quickly and in a peculiar manner, * * * twitching movements of toes begin in a few minutes, and soon extend to muscles of the leg and trunk. In some cases these movements were violent; strong convulsive contractions of muscles and limbs followed each other rapidly, and were often so violent as to throw the animal out of the position in which it was lying. The convulsions resembled those caused by strychnia, the violent tetanic contractions had the appearance of being caused by stimulation of the cord." This extremely excitable condition of the reflex nervous mechanism was followed after a time by the complete loss of irritability of this mechanism. These observations by Howell seem to me to show that between the stage of the normal irritability of this reflex mechanism, when the calcium salts are supplied to it in normal quantity, and the complete paralysis or loss of irritability of this mechanism, due to the more or less complete absence of calcium salts, which have gradually been washed away by the calcium free circulating fluid, there is a stage of extreme irritability, and reflex excitability of this reflex nervous apparatus, which corresponds to the period when this nervous mechanism is supplied with a diminished amount of calcium salts; that is to say, there is a partial calcium starvation of the nerve elements. This explanation of Howell's experiments is supported by his further experiments. In animals, in which the irritability of the reflex nervous apparatus had been destroyed by calcium starvation as in the above experiments, it was found, that if calcium solution was added to the circulating fluid of the muscle, the primary effect was to again produce a twitching movement of these muscles, "lasting for a short while," to be followed by a more or less distinct return of the muscle to its normal irritability. I have taken the liberty of drawing the above conclusions, which I believe to be correct, from the work of Howell and others, but to which I do not wish to commit Dr. Howell, as he made no such deductions from his experiments. The small amount of calcium, which first reached the muscle, resulted in a partial restoration of the nerve muscle irritability, and made possible the same convulsive movements which were above noted as being due to too little calcium in the circulating fluid, and these convulsive movements subsided when the nerve and muscle elements had received sufficient calcium to place them in a state of normal irritability. From these, and other experiments along the same line, I conclude that calcium starvation of nerve elements may be a factor in the production of the convulsive neuroses of childhood. The application of this conclusion to clinical medicine will, I believe, in the near future be recognized as something of real importance.

# VIII.

### REFLEX IRRITATION.

Reflex irritation is one of the most important etiological factors of the neuroses of childhood. Many able pediatrists in recent years have waged an active crusade against this proposition, which previously was thought to be one of the axioms of medical knowledge. While these men have not been able to convince the medical world that reflex irritation is an unimportant factor of neurotic disease, they have very much modified the view, which so long obtained, that reflex irritation was the all important factor in producing these diseases. In the proposition, as stated at the beginning of this chapter, I have taken position between these extreme views, and it will be the purpose of this chapter to show that the influence of reflex irritation in producing nervous diseases in childhood has been as much underrated in recent years as it was exaggerated by earlier writers, who taught that almost every nervous disease was caused by some reflex act. It is a matter of common clinical observation that such neuroses as hysteria, incontinence of urine, night terrors, chorea, convulsions, fever and headache are etiologically related to some form of reflex irritation, and this relationship is not infrequently absolutely demonstrated, when removal of the reflex irritation cures the neurosis.

The common sights of reflex irritations, which are recognized factors of nervous diseases in children, are the genito-urinary organs, the gastro-intestinal tract, the eye, the ear and the nose. The importance of this subject does not end with recognizing that reflex irritations from all of the above-named sights are common factors of neurotic disease, but it is of equal importance that we should recognize that, as a rule, reflex irritation acts conjointly with other factors in producing the neuroses of childhood. It is a well-known fact that reflex irritation, of apparently a severe type, may exist without producing nervous symptoms. In such instances, the center which is the most important part of the reflex arc is normal, stable and not easily excited to discharge its stored up nerve energy. It is most important, therefore, that we should recognize the fact that the reflex irritation, which excites neurotic disease, is made potent by reason of its connection with an abnormally irritable reflex center. In previous chapters we have studied the influence of heredity, sex, age, environment and various blood conditions, in producing an increased irritability of nerve centers; and it is chiefly with the aid of these factors of neurotic disease, that reflex irritation can produce such a wide range of nervous symptoms. The study of this subject embraces, therefore, not only how each of these factors may act in producing nervous symptoms in children, but it must also inquire into the inter-dependence and relationship of these factors in producing these symptoms.

The fact that reflex irritation is commonly associated with other factors does not in the least diminish its importance as a factor of neurotic disease, since the removal of the reflex excitant very commonly cures the neurosis, even though the other factors remain, and since our best efforts at removal of other factors of neurotic disease, as a rule, are futile for good, so long as the reflex excitant remains to constantly excite the nerve centers. The explanation of these clinical facts is, that reflex irritation does not act simply as an excitant in discharging nerve force from irritable centers, but it also acts by keeping up the irritabil-

ity of these centers, and, if long continued, by producing changes in the nerve centers recognizable
under the microscope, which make these centers
more irritable and more susceptible to reflex excitation.

If this be true, then, reflex irritation at once as-
sumes a commanding position among the factors of
neurotic disease in children; such a position as, in
recent years, has not been accorded to it, and it is
the special purpose of this chapter, to replace reflex
irritation in the high position which it merits
among the factors of neurotic disease in children;
in that position which it formerly occupied, and
from which it has been unjustly removed.

The microscope has gradually revealed to us the
fact, that all cellular activity is accompanied by
definite chemical and morphological changes in the
cell itself. The tired cell differs from the rested
cell, not only in morphological changes which can
readily be noted in nucleus and cell protoplasm,
but also in the reaction of both cell protoplasm and
nucleus to coloring matters.

The changes which result from the functional
activity of cells may be called fatigue changes,
and it is evident that the longer the cell is worked,
the more marked will be these changes. It is also

a physiological fact, that fatigue changes in the tired cell will disappear after a period of rest, and the cell will again be found morphologically and chemically a rested cell, but it requires a longer period of time for a cell to return to its rested condition than it does for the same cell to tire under ordinary work.

The fatigue changes, resulting from the functional activity of glandular epithelium are, as a rule, very pronounced. These changes, while not the same in all gland cells, may be noted in the shrunken condition of both nucleus and cell protoplasm and in the changed reactions to coloring matters of both nucleus and cell protoplasm. Fatigue changes in the tired muscle cell are also shown in the shrunken and vacuolated condition of its protoplasm. And both the tired muscle cell and the tired gland cell are only restored to their rested condition by a period of prolonged rest—the period of rest required being considerably longer than the period of activity.

The nerve cell, like the gland and muscle cell, shows marked morphological and chemical fatigue changes. C. F. Hodge, in a very clever piece of work, has shown that definite changes occur in the nerve cells of the brain and spinal ganglia of cer-

tain birds and bees as a result of their normal daily
activity. He compared the nerve cells of sparrows
and swallows, shot in the early morning, with
the nerve cells of sparrows and swallows, shot in
the evening, after a day of hard flight. Experiments of this kind on birds and bees invariably
showed fatigue changes in the nerve cells tired
from the day's work. Hodge also found definite
changes to occur in the spinal ganglion cells of the
frog, the cat and the dog, under electrical stimulation, and these changes were very similar to the
changes which he had observed to result from the
normal daily activity of nerve cells.

These fatigue changes in the nerve cells, whether
resulting from normal daily activity or electrical
excitation, are as follows :

Nucleus was " much smaller, and had a jagged,
irregular outline. It took a darker stain, and lost
its reticular appearance."

Cell protoplasm " did not take stain so readily,
and was much shrunken. In spinal ganglia it was
vacuolated."

Hodge also observed that the nerve cell recovered much more slowly than it tired, and that the
recovery of the nerve cell might be represented by
a curve, quite similar to the curves obtained by

Mosso and Lombard, for the muscle cell in its recovery from fatigue. He concludes that "individual nerve cells, after electrical excitation, recover if allowed to rest for a sufficient time, but
the process of recovery is slow. From five hours'
stimulation, recovery is scarcely complete after
twenty-four hours' rest."

The changes above noted in nerve cells, as resulting from electrical stimulation and normal fatigue, have a plain bearing on the study of the
changes which occur in the spinal ganglion from
reflex irritation, since reflex irritation can do nothing more than greatly exaggerate the functional
activity of these cells, and must, therefore, result in
changes within the cells similar to those above described.

Satovski, in a careful research on "Changes in
Nerve Cells Due to Peripheral Irritation," has made
an important advance in our knowledge of this
subject. He irritated a peripheral nerve by ligature, and thereby caused a peripheral, but not a
central, degeneration of the nerve. In this way,
he produced a chronic reflex irritation of that portion of the cord to which this nerve belonged, and
on microscopical examination of the cord, at this
point, he found on the injured side, using the unin-

jured side for a control, many cells exhibiting great vacuolation, and shrinking of the protoplasm from the capsule. The nuclei of these cells were oval instead of round, they stained easily, and were sometimes so much shrunken that they were zigzag in outline, and left a space between the protoplasm and the nucleus of the cell.

Mrs. Ternowski, in a research on "Changes in the Spinal Cord from Stretching the Sciatic Nerve," found changes very similar to those previously noted by Satovsky.

From the observations quoted, it is plainly evident that chronic reflex irritation can produce very marked changes in the nerve cells of the spinal ganglia, and that the longer and more violent this irritation is, the more pronounced will these changes be. It is also plain that a considerable length of time must be required to restore to their normal condition, cells which have been subjected to reflex irritation for months and years. It has even been noted that nerve cells, under electrical stimulation, can be so exhausted that the nuclei will entirely disappear, and the cells be unable to recover their normal condition, even after the removal of the stimulus which produced the change. Here we have an explanation of the ofttimes slow recovery

of an irritable spinal cord, after the removal of the
reflex cause which brought about the irritability.
In the application of these facts to clinical medicine, we must remember that the spinal cord has
but two functions, viz., conduction and reflex action.
We must also remember that a reflex irritation of
an afferent nerve, carrying impulses to any one of
the many special reflex centers of the cord, does
not confine its morbid influence to that center, but,
by reason of the physiological law of "overflow
of reflexes," the impulse spreads up and down the
cord, producing changes in the cells of adjacent
centers ; and if the reflex irritation be severe and
long continued, the impulses may spread throughout the cord involving all its centers, and producing a general spinal irritability, and in this way
predisposing the individual to all kinds of reflex
nervous diseases.

In some recent experiments made upon rabbits,
I have been enabled to demonstrate, that a chronic
reflex irritation can produce a most extreme irritability of the nervous centers in the cord of this animal. In these experiments, the abdominal cavity
of the rabbit was opened and the large intestine
stitched into the abdominal wound.   These rabbits
quickly recovered from the operation, and for a

week or ten days seemed normal in every way. At this time the reflexes, which in the normal rabbit can scarcely be brought out at all, began to be very perceptible. In these experiments the knee jerk, and a reflex, which is produced by letting the finger slip over the anterior superior spine of the ilium, were studied, and it was found that from the tenth day onward, there was an increase in the reflex excitability of the cord, as determined by an increase in the above-named reflexes. The reflex excitability of the cord continued to increase for about six weeks; after this period of time, the cord was so excitable that it was impossible to make out whether the excitability was increased or not, since a slight touch would produce a maximum reflex.

These experiments clearly show that chronic reflex irritation, unassisted by any other cause that could be made out by careful post-mortem examination, can produce in the rabbit a most extreme irritability of the spinal motor centers. The post-mortem examinations of these rabbits, one of which was killed three months after the operation, showed no evidence of peritonitis or other disease, other than the attachment of the large intestine to the abdominal wall. The spinal cord of the rabbit.

killed at the end of the third month, was examined microscopically, a number of sections being made from the lumbar and dorsal regions. In all of these sections changes in the ganglion cells, similar to those described by Satovski, were found. The nuclei were irregular in size and outline, many were oval and many had a jagged outline, many of the nuclei were small and had a shrunken appearance, and all of them took the stain more deeply than does the rested (normal) nucleus. The protoplasm of the cells did not take the stain as it normally does, and in many instances it took the stain so faintly that the outline of the cells could not be made out. In some instances only the small, deep-stained nuclei were visible.

In the above observations, we have not only a physiological, but also a morphological explanation, of how and why a chronic reflex excitation may be an important factor in producing a general spinal irritability, and we have also a sufficient explanation of the fact that the removal of the reflex cause, which has been acting for years in producing spinal irritability, may not at once be followed by the cure of the spinal irritability, and that it may even require years of comparative rest for the irritable spinal centers to become stable (normal), even after the removal of the reflex cause which produced the irritability of these centers; and these observations also justify the belief that reflex irritations, acute and chronic, are among the most important causes of neurotic disease in children.

In the study of the influence of reflex causes in producing the neuroses of childhood, one important question must be answered, viz.: Why is it that chronic reflex irritation is so much more important, as a factor, in producing nervous diseases in children, and in girls, than it is in men? The oculist will testify that eye-strain is a much more potent factor in producing headache, chorea and general nervous irritability in children, and in young women, than it is in men. The surgeon will testify that diseases of the genito-urinary apparatus, which produce the most profound nervous symptoms in women and children, have little or no such influence in men. The physician will testify that irritation from disease of, or foreign bodies in, the intestinal tract will produce convulsive and other nervous disorders in children, while the same conditions have little influence in producing nervous symptoms in men. The gynecologist is prone to believe that disease of the female generative organs

is the most important of all the reflex causes of
nervous disease; and every clinician has observed
the predisposition to nervous disease, which accompanies the growth and functional development of
these organs. In fact every department of medical
science lends testimony to the fact, that age and
sex are among the most important of the predisposing factors, which assist reflex irritation in producing neurotic disease; and the reasons for the
potency of reflex causes in producing neurotic
disease in children and girls are not altogether obscure.

The following facts may be noted :

1st. **In children.** (*a*) Reflex causes are more
frequent than in adults, such for example as uncor-
rected eye-strain, adherent prepuce, balinitis, etc.
(*b*) The nervous system of the child is more irritable
and unstable by reason of its incomplete functional
development. (*c*) The inhibitory control of higher
nerve centers on spinal reflex movement is feebly
developed in the child. (*d*) Blood changes are
much more common allies of reflex disturbances,
in producing nervous disease, in children than they
are in adults.

2d. **In girls.** (*a*) Reflex causes are very much
more frequent than in boys or in female adults,

(the approach of puberty, with the functional development of ovaries and uterus, is a source of constant reflex disturbance; after the full functional development of these organs, the reflex excitation is intermittent and confined to a period just before and during a menstrual period). (*b*) Inhibitory control of the spinal motor centers more readily gives way in young girls than in boys of the same age. (*c*) The social conditions and habits of life of the young girl predispose her to nervous disease. (*d*) Blood changes, which produce nervous irritability, are very much more common in girls than in boys.

The above are some of the factors which assist reflex causes in producing neurotic disease in children and in young girls, but which have little influence in producing disease in male adults. In these observations we have an answer to the question : Why does reflex irritation produce nervous disease more readily in the child and young girl, than it does in the male adult?

# IX.

## EXCESSIVE NERVE ACTIVITY.

There is a well-grounded and wide-spread medical opinion that excessive nerve activity, (the term including brain work and nerve excitement) is an important factor in the production of nervous disease in children, but notwithstanding the prevalence of this belief among medical men, very little has been done to educate those, who have the rearing and tutorage of the young, on this subject, which, I believe, is one of almost vital importance to the state itself.

It is a fact which should be heralded every-where that the vast army of neurasthenics and hysterics, which now inhabit our cities, is yearly being increased by subjecting the immature nervous systems of young children to the almost constant excitement, strain and mental activity with which our social order has surrounded them. An all important question, therefore, to pediatrists, who should be especially interested in making of the child the strongest possible man, is : How can these influences, which are playing such havoc with the nervous systems of children, be guarded against? How can they be counteracted? How can parents, guardians, nurses and teachers be made to comprehend the importance of this subject?

If these questions are to be answered ; if the campaign against the evil of constantly subjecting children to the nervous strain, resulting from the artificial conditions which obtain in all cities, is to be, in any degree, successful, then the whole subject must be placed upon a more exact physiological basis than it has ever been before, so that those who have charge of the young may be told not only that nervous strain is an important cause of neurotic disease, but they may also be told why this is so. And in the series of papers which this chapter concludes, I have attempted to outline some of the physiological facts by which this goal is to be approached.

The teachers and guardians of the young must be told that the nervous system of the child differs very materially from the nervous system of the adult; they must be told that the child, especially in his nervous organization, is not a *little man;* that his nervous system is structurally and functionally immature; that it is excitable, unstable and under feeble inhibitory control; that the sources of reflex irritation in the child are many, and that the nerve centers discharge their force more fitfully and readily than in the adult; that the period corresponding with the onset and establishment of the reproductive function in girls, is a time when they are especially predisposed to nervous disease. And they must also be told that these, and other physiological peculiarities of the nervous system of childhood, are made much more potent for evil when they are associated with the various "blood conditions," which in previous chapters, I have shown to be etiologically related to the neuroses of childhood.

In order to approach this subject in a physiological way, I shall call attention to a recent very extensive research by Dr. W. Townsend Porter, which has, I believe, great practical importance in the study of the influences of school life in producing the neuroses of childhood.

Dr. Porter demonstrated that children who are advanced in their studies, are, on the average, heavier, taller and of larger girth of chest, than less advanced children of the same age. Thus, boys aged eleven, were found in Grades I, II, III,

IV, V and VI, of the St. Louis Public Schools.
The average weight of the four classes was re-
spectively 64, 66, 68, 71, 72 and 74 pounds. The
ability to succeed in school life is, in the average, a
measure of mental power, and if successful scholars
are, as a rule, better developed physically than the
less successful, it follows that mental ability is, in
the average, greater in large children than in small
children of the same age.

Dr. Porter makes a practical deduction from the
law thus established. The entrance to any grade
in a graded school system is guarded by examination, and the children found in that grade are such
as have passed the entrance examination, and have,
in this way, shown their capacity to do the mental
labor exacted in this grade. The greater number
of these children are of the same age. The work
of this grade is, then, normal for this age, and the
average height, weight and girth of chest of this
age, form the physical development most often
found in children able to do the work of the grade.
No child younger than the average age of any
grade should be permitted to enter it, until a physical examination has shown that his strength shall
probably be sufficient. In determining this, the
relation of weight and girth of chest to height is

of special importance. Abnormal height is un-
doubtedly a disadvantage, yet such children may
be strong, provided their physical development is
in proportion to their height. If the contrary is
the case, the child will be much less able to resist
the strain of school life.

Dr. Porter points out the importance of frequent
weighings of growing children. Persistent loss of
weight in an adult is a matter of grave concern.
The failure of a child to make the normal gain in
weight is no less grave, and should lead to an in-
quiry into his school tasks, for the effects of prolonged overwork are very serious in children, and
often irremediable.

It is my belief that if there was a rule, such as
Dr. Porter suggests, guarding every grade in our
public school system by a physical as well as a
mental examination, it would prevent the development of a considerable portion of the neurotic
disease, which is now so prevalent among school
children. With children of good physical development, working in the public schools within the
limitations of their proper grades, there is almost
no danger that a moderate amount of school work
will in any way assist the development of neurotic
disease, provided always that the hygienic condi-

tions of the school, especially the light and venti-
lation, are good. But the strain of ordinary school
work is a very different matter with children of
poor physical development, many of whom are,
unfortunately, precocious. A large number of these
children, by reason of bad heredity, are neurotic,
poorly nourished and anemic, and many of them
have tuberculous, rheumatic or syphilitic inheri-
tance, while others, from accidental causes, such as
bad hygiene, improper food, etc., are below the
normal in physical development. The nervous
systems of such children are in a condition of
malnutrition, and are, therefore, not capable of
doing the ordinary work of their grades in the
public schools, and if they are permitted to do
this work, or if, as is often the case, these children are encouraged to push on into higher grades
than the one to which their years and strength
should assign them, disastrous consequences will
surely follow, and their nervous systems may be
injured beyond repair.

These children, under the mental strain of school
work, may develop chorea, hysteria and other
neuroses. The important duty, therefore, of every
physician is to advise against much school work in
children of feeble physical development, and to

explain to parents and teachers why such children
as these should first have their physical defects
looked after, and should then be placed in a grade
lower than that to which their age and intelligence
should assign them.

It is my belief that a normal dwarf, with no bad
hereditary influences behind him, may, without injury to himself, keep pace in mental development
with fellows of his own age; the dwarfish body is
not of itself an indication that school work might
be injurious, if there is every other evidence of
perfect physical development.    Dwarfishness of
body in school children of good physique does not
mean dwarfishness of mind.    But dwarfishness
among children, as indicated by weight and chest
development, is, as a rule, the result of disease and
bad heredity, and this is the reason why children
who are under weight and have poor chest development, are, as a rule, incapable, without injury to
their nervous systems, of doing the same amount
of school work as their fellows of the same age.
It is my belief, therefore, that the physical basis of
precocity and dullness in children depends upon
the facts that bad heredity and disease are the
chief causes of abnormal dwarfishness or poor
physical development in the young.    It is also my

belief that children of this class are, as a rule, anemic and poorly nourished, and that their nervous systems are therefore in a condition of malnutrition, and not capable of doing an amount of work in keeping with the age of the child.

The reasons, then, are clear why we should not allow a child of poor physical development to be pushed to rapid brain development. If we do, their nervous systems will surely suffer from the strain, and whatever predisposition they may have to neurotic disease will be greatly increased. In dealing with individual cases, it will be of the utmost importance to the physician to know the child's heredity; if the child has a bad family history, it should be the imperative duty of the physician to protect it against mental overwork. We can not, of course, change the child's ancestry, but we can speak out against the crime of pushing children with hereditary physical defects to rapid brain development, and in this way *developing an hereditary or acquired nervous weakness into actual disease.* School work may therefore be classed as a cause of neurotic disease in children of poor physical development, and it acts chiefly in calling out hereditary defects of the nervous system. In speaking of the school work as a cause of neurotic

disease in children, it must be understood that this
term embraces not only brain work, but also the
mental excitement which attends examinations,
and the eye strain which results from imperfect
vision and bad light, the latter being one of the
most common causes of reflex nervous disease in
children, and one of the physical defects which
should be promptly removed.

It must be remembered that what is here said of
the physical basis of precocity and dullness is a
matter of proof, and not of opinion, and that it ap-
plies to children only, and has nothing whatever to
do with the question of whether, in adult life, a
healthy body adds strength and capacity to the
nervous system.

In this demonstration of the injury which results
to the nervous system of the delicate child from the
nervous strain of school life, we have a most important warning against the pernicious habit of en-
couraging mental precocity in early childhood. It
is a matter of almost daily experience to see a
poorly nourished tuberculous child brought forward
for the purpose of demonstrating its "wonderful"
precocity. The proud mother and over zealous
nurse commence the process of mental cramming

even before infancy has passed into childhood. From this time on, children are daily being taught, apparently with the idea of destroying their childhood, and making of them little men and women. And this unphysiological process is not infrequently a factor in the production of the nervous disorders of late childhood, puberty and adult life. Mothers must be told that *early precocity is an abnormal condition in the human infant*, which, if encouraged, may result in actual disease and permanent mental impairment. They must be told that *vegetation* is the ideal life of infancy and early childhood. Look to the physical, and retard the intellectual development of the child. It must not be taught, it must not be trained. It must have plenty of exercise, fresh air, proper food and, if possible, a large portion of the year should be spent in the country, away from the clamor and excitement of city life. In the country, also, the child can have a certain amount of solitude, the importance of which can scarcely be overestimated in giving independence of thought and character to the future man.

It is my belief that the nurse and the governess in the modern home are doing much to destroy the development of individuality in children. The

modern child has some one to do his thinking, some
one to minister to his every want, and is almost
constantly being trained. He has no time to himself, and a very small portion of his day is spent in
play with his intellectual equals. If there is one
crying evil common to all of our large cities it is
the absence of play-grounds for children, and the
attention of humanitarians should be called to this
fact. If our generous citizens would pause long
enough in the building of hospitals, libraries and
places of learning, to realize that there is a field almost totally neglected by the humanitarian, and
one of as much importance to the welfare of our
communities as the building of hospitals, libraries
and institutions of learning; then possibly a portion of the vast sums of money annually spent in
this way would be spent in providing play-grounds
for children. These play-grounds should not be
covered with beautiful grass plots, guarded by po-
licemen, but they should be play-grounds in the
best sense of these words—places where ball, tennis,
and all kinds of healthful sport could be enjoyed.
And I believe the day is not distant when the
physiological importance of the physical, as opposed to the mental development of children, will

be so generally recognized that some philanthro-
pists will prefer to hand their names to posterity
associated with " play-grounds," rather than with
fountains, art museums, music halls and other
worthy enterprises.

# INDEX.

(119)

www.ingramcontent.com/pod-product-compliance
Lightning Source LLC
Chambersburg PA
CBHW021936190326
41519CB00009B/1032